LIGHT SPECTROSCOPY

The INTRODUCTION TO BIOTECHNIQUES series

Editor:

D. Billington School of Biomolecular Sciences, Liverpool John
Moores University, Byrom Street, Liverpool L3 3AF

Series adviser:

P.M. Gilmartin Centre for Plant Biochemistry and Biotechnology,
University of Leeds, Leeds LS2 9JT

CENTRIFUGATION
RADIOISOTOPES
LIGHT MICROSCOPY
ANIMAL CELL CULTURE
GEL ELECTROPHORESIS: PROTEINS
PCR
MICROBIAL CULTURE
ANTIBODY TECHNOLOGY
GENE TECHNOLOGY
LIPID ANALYSIS
GEL ELECTROPHORESIS: NUCLEIC ACIDS
LIGHT SPECTROSCOPY

Forthcoming titles

PLANT CELL CULTURE
MEMBRANE ANALYSIS

LIGHT SPECTROSCOPY

David A. Harris

Department of Biochemistry, University of Oxford, Oxford, UK
and
St Anne's College, Oxford, UK

Taylor & Francis
Taylor & Francis Group

LONDON AND NEW YORK

First published 1996

Transferred to Digital Printing 2006

A CIP catalogue record for this book is available from the British Library.

ISBN 1 872748 34 1

Published by Taylor & Francis
2 Park Square, Milton Park, Abingdon, Oxon, OX14 4RN
270 Madison Ave, New York NY 10016

DISTRIBUTORS

Australia and New Zealand
 DA Information Services
 648 Whitehorse Road, Mitcham
 Victoria 3132

India
 Viva Books Private Limited
 4325/3 Ansari Road
 Daryaganj
 New Delhi 110002

Singapore and South East Asia
 Toppan Company (S) PTE Ltd
 38 Liu Fang Road, Jurong
 Singapore 2262

USA and Canada
 Books International Inc.
 PO Box 605
 Herndon, VA 20172-0605

Typeset by Chandos Electronic Publishing, Stanton Harcourt, UK.

Contents

Abbreviations

9-AA	9-aminoacridine
ABTS	2,2'-azino-di(3-ethylbenzthiazoline)-6-sulfonate
ACMA	9-amino-6-chloro-2-methoxyacridine
A_λ	absorbance at wavelength λ
AM	acetoxymethyl (ester)
AMC	7-amino 4-methyl coumarin
ANS	1-anilinonaphthalene 8-sulfonic acid
ATP	adenosine 5'-triphosphate
BCECF	2',7'-*bis*-(2-carboxyethyl)-5-carboxyfluorescein
Dansyl	5-dimethylamino naphthalene 1-sulfonyl
DEPC	diethyl pyrocarbonate
DTNB	dithio-*bis*-nitrobenzoic acid
EDTA	ethylenediamine tetraacetic acid
EGTA	ethyleneglycol *bis*(β-aminoethyl ether) N,N,N',N'-tetraacetic acid
ε_λ	molar absorptivity at wavelength λ
F_λ	fluorescence at wavelength λ
FITC	fluorescein isothiocyanate
fluo-3	1-[2-amino-5-(2,7-dichloro-6-hydroxy-3-oxo-3H-xanthen-9-yl)] -2-(2'-amino-5'-methylphenoxy)ethane N,N,N',N'-tetraacetic acid
FRET	fluorescence resonance energy transfer
fura-2	1-[2-(5-carboxyoxazol-2-yl)-6-aminobenzofuran-5-oxy]-2-(2'-amino-2'-methylphenoxy)-ethane-N,N,N',N'-tetraacetic acid
GAP	glyceraldehyde 3-phosphate
GOD	glucose oxidase
HEPES	4-(2-hydroxyethyl)-1-piperazine ethanesulfonic acid
Hoechst 33258 (H33258)	2-[2-(4-hydroxyphenol)-6-benzimidazolyl]-6-(1-methyl-4-piperazyl) benzimidazole
hplc	high performance (high pressure) liquid chromatography
i.d.	internal diameter
indo-1	1-[2-amino-5-(6-carboxyindol-2- yl)phenoxy]-2-(2'-amino-5'-methylphenoxy)ethane N,N,N',N'-tetraacetic acid
LED	light-emitting diode
MOPS	morpholino sulfonic acid
NADH	nicotinamide adenine dinucleotide (reduced form)
Nbf	7-nitrobenzofurazan
OD	optical density
oxonol V	*bis*(3-phenyl-5-oxoisoxazol-4-yl)pentamethineoxonol
PEP	phosphoenol pyruvate
POD	peroxidase
Q	quantum yield
SEM	standard error of the mean
SNARF	10-amino-3-hydroxy-spiro[7H-benzo(*c*)xanthene-7,1'(3'H)-isobenzofuran]-3'one
TNBS	2,4,6 trinitrobenzene sulfonate
UV	ultraviolet

Preface

The spectrophotometer on the bench is a standard feature of all biochemistry laboratories. It requires no great skill to use, is sensitive enough to handle materials at physiological concentrations, and best of all, it produces immediate data. Often, however, this familiarity can lead to contempt – too many spectrophotometers are used without sufficient care, are rarely serviced or independently calibrated, and often, are underused or even misused in terms of the facilities they provide. With the current generation of instruments being increasingly automated, and data collection computerized, the gap between the user and the instrument widens – and this too can lead to misuse or misinterpretation.

This text aims to provide a guide to the principles behind light absorption and fluorescence and the quantitative measurement of these phenomena. Chapter 1 considers the interaction of light with matter, why it may or may not be absorbed or scattered, and if it is absorbed, what might happen to the energy concerned. Having thus established that measurement of absorption and fluorescence might yield useful information about biochemical systems, Chapter 2 investigates the types of molecule that, in biological systems, can be responsible for the absorption of light and introduces the concept of a probe, a nonbiological chromophore which can signal useful information about systems into which they are introduced.

Chapter 3 then outlines the basic components and construction of a spectrophotometer and fluorometer, a theme which is expanded in Chapter 4, where the details of their design are amplified to explain the approaches taken to maximize accuracy and precision of measurement. The principles behind single beam, dual beam, and dual wavelength instruments are explained, with reference to commercially available instruments. The complications involved in measuring and quantitating fluorescence are also discussed. Chapter 5 deals with methods of preparing samples in order to optimize measurements.

The remainder of the text deals with some applications of the measurement of absorption and fluorescence. Chapter 6 covers the measurement of absorption spectra in both optically clear and turbid samples, and how these spectra are used for identification of individual compounds and, hence, for the characterization of chemical changes such as electron flow in mitochondria, photooxidation in chloroplasts and the modification of protein side chains. Fluorescence spectra are covered in Chapter 7. More quantitative aspects of analysis are discussed in Chapter 8, which deals with assays – of pure compounds, of individual compounds in complex mixtures and of rates of processes such as enzyme-catalyzed reactions. This chapter attempts not only to cover the methods but also to entertain considerations of sensitivity and assay precision and, in particular, how far the results of an assay may be believed.

Chapter 9 includes a variety of techniques, linked together by their employing spectroscopic methods to probe aspects of biological systems. This includes measurements of parameters as disparate as ligand affinities for proteins, membrane potentials, intracellular ionic concentrations and intramolecular distances. The list is by no means exclusive – the intention is to draw attention to the potentials, and limitations, of the methods rather than to provide a comprehensive overview. Hopefully, the reader will be inspired to develop his own methods based on those outlined here.

I should like to thank Kontron Instruments for their help in providing optical diagrams, to Hi-Tech Scientific, Thorn-EMI Electron Tubes Ltd and Dr S. Colloms (Microbiology Unit, Department of Biochemistry, University of Oxford) for providing photographs and to Mr Matthew Rowe for his artistic and image manipulation skills in the design of the cover picture. I am also grateful to Ms Sabine Benz for providing several of the figures and Ms J. Harrington for assistance in typing the text.

David A. Harris

1 Principles of Spectrophotometry

1.1 The nature of spectrophotometry

Spectrophotometry, as used in the life sciences, involves measuring light absorbed by compounds in (generally aqueous) solutions. We can obtain information both from the amount of light absorbed, and from the wavelength(s) at which absorption occurs.

Spectrofluorometry again deals with compounds in solution, but we now measure light emitted by such compounds while they are, or immediately after they have been, illuminated. Again, information is available from the amount of light emitted, and its wavelength.

Both terms relate to the interactions of matter with electromagnetic radiation of wavelengths between about 200–1000 nm (*Figure 1.1*). This includes visible light (350 nm blue to 700 nm red), and also the near ultraviolet (200–350 nm) and near infrared (700–1000 nm). These are the wavelengths with which we are most familiar (through vision etc.), and, not wholly unrelated, include those wavelengths which most readily pass through the earth's atmosphere. In addition, such radiation does not readily damage biological molecules (although damage is beginning to occur in the ultraviolet region); thus, unlike techniques employing, say, X-ray radiation, spectrophotometry and fluorometry can be used as nondestructive techniques for probing biological systems.

Another feature of this range of wavelengths is the type of molecular processes they trigger. The energy of this radiation is such that it induces transitions in the outermost electron shell of the molecules under study. Thus we observe electrons moving between one molecular orbital (e.g. π) and another (e.g. π^*), rather than changes in

1

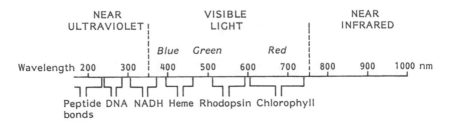

FIGURE 1.1: *Region of the electromagnetic spectrum in this text.*

bond vibrations or in atomic nuclei. This will become clearer after considering the nature of light and how it interacts with matter.

1.2 Interaction of light with matter

Like any radiation, light has a dual nature – as an electromagnetic wave and as a stream of particles (photons). If we consider its nature as an oscillating electromagnetic field, we can see how light can interact with electrons (and their associated electric fields) in atoms and molecules. If we consider light as a stream of photons, we can see how it can provide precise amounts (quanta) of energy to such electrons. The two aspects merge when we calculate the amount of energy per photon, *E*, which is given by

$$E = hc/\lambda = h\nu$$

where λ is the wavelength of the light, ν its frequency, **c** the speed of light *in vacuo* and h = Planck's constant. From this equation, we can calculate that one photon of yellow light (λ = 500 nm) provides 4×10^{-22} J; expressing this in more accessible units, we calculate that 1 mol of photons (6×10^{23} photons; 1 einstein) carries 240 kJ.

When a beam of light is passed through a solution, several things might happen (*Figure 1.2*). The simplest possibility is that the beam emerges in the same direction with the same intensity (transmission) (*Figure 1.2a*). On its passage, the electromagnetic wave will have set up transient oscillations in the electrons of the molecules of the solution, and hence it will have been slowed down in transit (refraction). Nonetheless, these oscillations remove energy from, and reinforce, the traversing wave equally and, in principle, no intensity is lost.

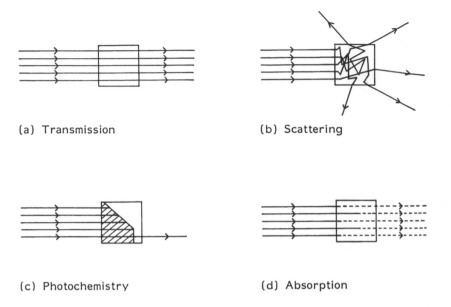

(a) Transmission

(b) Scattering

(c) Photochemistry

(d) Absorption

FIGURE 1.2:   *Interaction of light with material in solution.*

In fact, some intensity is always lost because some light is always diverted to emerge in a direction different from that of the incident beam (*Figure 1.2b*); this is scattered light. All wavelengths of the incident light are scattered, to nearly (but not precisely) the same extent (see Section 1.8). For a true solution, only a small fraction of the incident light is scattered, and the solution is said to be optically clear. If, however, the solution contains particles larger than the wavelength of the incident light (>1 µm) (i.e. it is a suspension rather than a true solution), much more light is scattered. This can be observed, for example, if a cup of tea is observed before (optically clear) and after adding a suspension of fat globules (milk). To put this in a biological context, bacteria are typically about 10 µm across and bacterial suspensions scatter light considerably; viruses are typically 50–100 nm across, and solutions of viruses transmit light well (*Figure 1.3*).

A third possibility is that light induces some chemical change in the molecules in solution – for example by providing enough energy to completely expel an electron from a molecule. This creates a highly reactive free radical, which then undergoes further chemical reactions (*Figure 1.2c*). While this is unusual within the energy range of visible light, some biological systems (such as plant chloroplasts) are adapted to use visible light to drive biochemical processes in this way. As seen above, the amounts of energy carried by photons of visible light are physiologically useful – our einstein of yellow light described above carried 240 kJ of energy, which is comparable with the 60 kJ required to synthesize 1 mol of adenosine triphosphate (ATP). As we move into

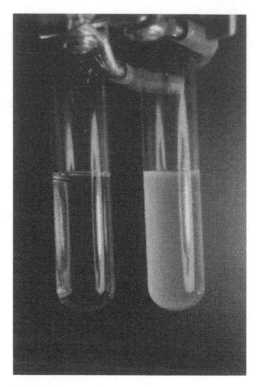

FIGURE 1.3:   *Growth medium containing 10^8 virus particles (left hand tube) or 10^8 bacteria (right hand tube) per ml. Photograph kindly provided by Dr S. Colloms.*

the ultraviolet region (especially below 260 nm), light induced photochemistry can have a deleterious effect on proteins and nucleic acids. However, a detailed discussion of these processes is outside the scope of this text.

We are particularly concerned with the final possibility, that of the absorption of light by molecules. If the wavelength of the incident light is such that the energy per photon precisely matches the energy required to excite an electron (in one of the molecules in solution) to an orbital of higher energy, a specific interaction may occur between the photon and the electron, and this light does not emerge; it is absorbed in the solution (*Figure 1.2d*). Since this is a specific interaction – it requires the energy of the photon to match the energy of the electronic transition — only certain wavelengths will be absorbed. It is this variation in absorption with wavelength which makes the phenomenon of light absorption such a valuable tool in characterizing and quantitating biological compounds.

1.3 Absorption lines and absorption bands

Electrons in a molecule normally occupy that set of molecular orbitals of lowest energy, that is, electrons are distributed among the bonds so that the molecule is in its most stable electronic configuration. This is called the electronic ground state of the molecule. It is possible that one of the electrons is found in an orbital of higher energy, and the molecule is in an excited state (*Figure 1.4*); however, at room temperature, this is very unlikely and only 1 in 10^6 molecules are typically found in an electronic excited state.

This is because a considerable amount of energy is required to promote an electron from the ground state arrangement to an excited state – much more than the molecule is likely to gather by random collisions with other molecules. However, this amount of energy is comparable with the amount carried by a photon of light. If the wavelength of the light is such that the energy per photon precisely matches the energy difference between the two electronic energy levels, the photon is absorbed and the electron is promoted to its upper energy level. If the solution is illuminated with a range of wavelengths, light at only one particular wavelength is absorbed, and we see an absorption line (*Figure 1.5a*).

In fact, absorption lines are observed only with dilute, atomic, gaseous samples. In molecular solutions, absorption occurs over a range of wavelengths (typically 10–40 nm wide) around that corresponding to an electronic transition; we see an absorption band rather than a sharp line. This is because a molecule in solution contains several atoms in motion relative to each other, and is continually buffeted by solvent molecules with which it can exchange energy. This has two effects. Firstly, not all molecules have precisely the same energy at a

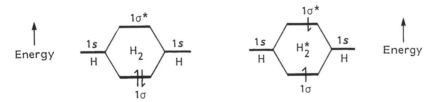

Ground state of H_2 molecule   1st excited state of H_2 molecule

FIGURE 1.4: *Electronic states of the hydrogen molecule. σ represents a molecular orbital, here derived from the parent (atomic) s orbitals.*

given instant; atoms are vibrating or rotating differently relative to each other. Since the amount of energy per quantum for vibrations and rotations is relatively small, a significant fraction of molecules lie above the ground state for such motions (represented by the thinner lines in *Figure 1.5b*). This gives rise to 'fine structure' around the main absorption line as molecules can be excited from a range of electronic ground states to a range of electronic excited states, the range being set by the vibrational and rotational energy levels. Secondly, as collisions between the absorbing molecule occur very frequently (>10^{13} times per sec in solution), the amount of time a particular molecule spends in a particular energy level (the lifetime of given state) is very short. Heisenberg's Uncertainty Principle then leads to a blurring of the spacing between energy states and a broad range of wavelengths absorbed (*Figure 1.5c*).

1.4 Absorption spectra

We have seen how a given compound will preferentially absorb certain wavelengths of light, dependent on the electronic structure of its molecule. Bands of specific wavelengths may be removed by

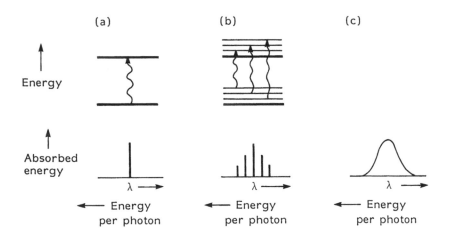

FIGURE 1.5:   *Absorption of energy as a function of wavelength. (a) For a transition between two electronic energy levels. (b) For a molecule with additional vibrational/rotational energy levels, (note that the lowest energy state of the molecule includes some vibrational/rotational energy). (c) The theoretical line spectrum (b) broadened by intermolecular collisions in solution.*

absorption. If this occurs, the solution appears colored. For example, if we shine white light through a solution of chlorophyll, red and blue light is preferentially absorbed; green light is transmitted, making the solution appear green (*Figure 1.6*). A different compound will absorb at different wavelengths; hemoglobin absorbs green light and its solution appears red. Thus compounds can be identified or characterized by measuring absorption as a function of wavelength – an absorption spectrum.

Figure 1.7 shows the absorption spectrum of chlorophyll. The absorption shows a pattern of maxima (peaks) and minima (troughs) (where a tangent to the curve is parallel to the abscissa). As we noted above, chlorophyll has a strong absorption peak in the blue and another in the red, but absorbs little green light. A compound is characterized by the position of, in particular, the peaks; the wavelength corresponding to a peak is denoted by the symbol λ_{max}. Differences may be more subtle, but still detectable. Hemes *a,b* and *c* (in their respective cytochromes) give similar but clearly distinguishable spectra (see Chapter 6).

Finally, since an absorption spectrum is related to chemical nature, it will change if one chemical species is converted into another. This may occur, for example, by oxidation or reduction. *Figure 1.8* shows the changes in an absorption spectrum occurring as cytochrome *c* is reduced. Peaks at 418 nm (γ), 522 nm (β), and 550 nm (α), appear on reduction. Changes in the absorption spectrum can thus be used to monitor the progress of the reduction. A further feature to note is the point(s) at which these spectra cross, which represent wavelength(s) at which absorption is unaltered in passing from the oxidized to the reduced form. These are called isosbestic points. In any two-state system (oxidized → reduced; folded → unfolded), there must be at least one isosbestic point in the two spectra, and practically this provides a valuable reference point when measuring spectral changes in complex or turbid solutions (see Chapter 6).

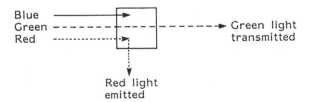

FIGURE 1.6: *Solutions of chlorophyll appear green, as they absorb blue and red light. Some of the absorbed (red) light can be re-emitted as red fluorescence.*

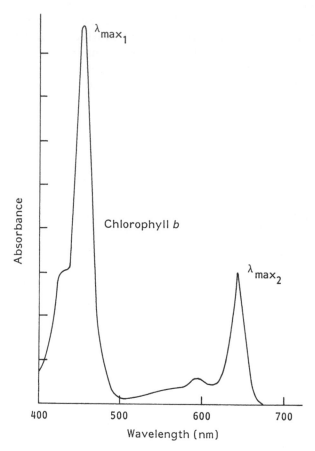

FIGURE 1.7:   *Absorption spectrum of chlorophyll* b.

1.5 Quantitative aspects of absorption

Photons, impinging on a solution of absorbing material, may either pass through (transmission) or be absorbed. As we have seen above, the chance of absorption depends on the type of molecule and wavelength associated with the photon (i.e. on its color). However, absorption also becomes more likely as the number of molecules in the light path increases; this in turn depends upon (i) the length (l) of the light path and, (ii) the concentration (c) of absorbing molecules. Since the chance of absorption is the same for each individual photon, this leads to an exponential decrease in the number of photons transmitted as they cross the solution. In other words, if half the

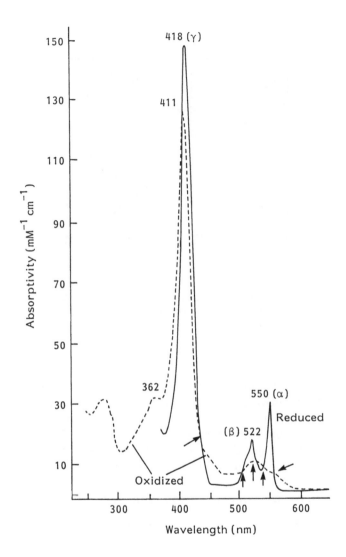

FIGURE 1.8: *Visible absorption spectra of reduced (solid line) and oxidized (dashed line) cytochrome* c. *The isosbestic points are indicated by arrows. Data from Meyer and Kamen (1982) Adv. Prot. Chem.,* **35,** *105–212.*

photons are absorbed in d cm, then half of the rest are absorbed over the next d cm, half of the rest over the next d cm and so on. This is shown in *Figure 1.9a*.

Mathematically this is represented by

$$N = N_o \exp(-kcl) \quad \text{or} \qquad (1.1a)$$
$$I = I_o \exp(-kcl) \qquad (1.1b)$$

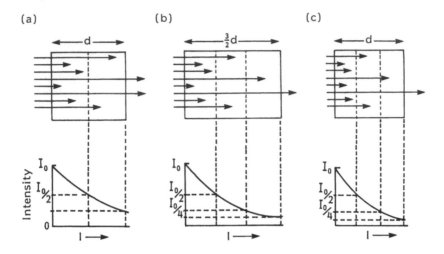

FIGURE 1.9:   *Passage of light through a solution of absorbing material. (a)
Half the light is absorbed over half the cell length, d. (b) The effects of
increasing cell length. (c) The effect of increasing the chance of absorption
per unit length (via increasing c or k).*

where N = number of photons and I = intensity of the light beam at a
distance l, and N_0 and I_0 the respective initial values (when $l = 0$). k is
the 'decay constant', which is dependent on the type of molecule
present. The effects of varying c and k on light intensity are indicated
in *Figures 1.9b* and *1.9c* respectively; if k is large, the intensity
declines rapidly and the compound is said to be strongly absorbing.

From the above equations, we can define two parameters:

$$\text{transmittance } (T) = I/I_0 = \exp\,(-kcl) \text{ and} \qquad (1.2)$$
$$\text{absorbance } (A) = -\log\,(I/I_0)\ = +(2.3k)\,cl \qquad (1.3)$$

From Equation 1.3 we can immediately see that absorbance is a
useful parameter, being linearly related to concentration. Indeed,
absorbance is widely used in biochemistry and biology to measure
concentrations. The absorbance of a solution when $l = 1$ cm (a common
pathlength in laboratory instruments) may be termed the optical
density (OD) of the solution.

Equation 1.3 can be rewritten as:

$$A = \varepsilon cl \qquad (1.4)$$

where the composite proportionality constant, $2.3k$, is replaced by the
symbol ε, which again is dependent on the type of molecule absorbing.

Equation 1.4 is known as the Beer–Lambert Law (or, commonly, as Beer's Law). Comparing Equations 1.2 and 1.4, we see that

$$A = -\log (I/I_0) = -\log T \qquad (1.5)$$

Thus an absorbance of 1 indicates a transmittance of 10% (only 10% of the incident light passes through), and absorbance of 2 indicates a transmittance of 1%, and so forth. Measuring absorbances above 2 or 3 requires a spectrophotometer of specialized design, because the amount of light passing through the sample, and hence available to be measured, is so small.

When concentration is given in mol l^{-1} (M), and l in cm, ε is termed the molar absorptivity (sometimes called the extinction coefficient), and represents the absorbance of a 1 M solution in a 1 cm light path. For macromolecules like DNA, where the molecular weight may be unknown or heterogeneous, an alternative convention is to express the concentration in percent (a 1% solution contains 10 mg solute per ml), giving specific absorbtivity ($\varepsilon^{1\%}$). ε is typically recorded at an absorption maximum (λ_{max}), which may be given as a subscript (ε_λ). Typical values of ε for some common biochemicals are given in *Table 1.1*.

In some measurements, the Beer–Lambert Law appears to break down, and absorbance loses its proportionality to concentration – typically at high concentration. On standard laboratory spectrophotometers, this may occur due to instrumental limitations and the difficulties of measuring low levels of transmitted light, in particular due to problems with 'stray light' (see Chapter 4, Section 4.1). However, true deviations may also occur, which indicate changes in the chemical species present in solution – for example aggregation or ionization at high concentrations, which may perturb the electronic structure of the molecules in solution.

1.6 Light emission in solution – fluorescence

After it has absorbed light energy, a molecule has an unstable electronic configuration. The molecule therefore loses this energy very rapidly (typically in <1 nsec) as it regains its most stable configuration. In most cases, this energy is lost as heat, that is, it is converted into molecular rotations and vibrations. This occurs because there is a continuum of rotational and vibrational energy levels in the ground electronic state which stretch up to the energy of the excited state (see *Figure 1.10*). Although these energy levels are

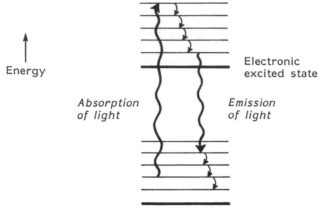

Energy

Absorption
of light

Emission
of light

Electronic
excited state

Electronic ground state

FIGURE 1.10:   *Loss of energy from electronic excited state of a molecule. Energy is lost as heat (arrows) through vibration or rotation, and as light (bold arrows) by fluorescence emission.*

normally empty, they provide a facile route for an excited molecule to dispose of its excess energy. Transitions between these states occur on the same time scale as molecular collisions in solution ($\sim 10^{13}$ times per sec) and thus the excited molecule has a lifetime of less than 10^{-12} sec.

In some molecules, for example rigid ring systems, rotations and vibrations may be too restricted to allow the excited state to dissipate its energy as heat. In this case the excited state may re-emit the absorbed energy as light, in a single de-excitation event. This process is called fluorescence emission. Emission of such a large packet of energy (compared with rotational and vibrational quanta) is less frequent, and the excited state molecules survive for rather longer – about 10^{-9} sec. This is the fluorescence lifetime of the molecule.

Because of the energies involved, the wavelength of the light emitted is similar to that of the wavelength absorbed. Thus a solution of chlorophyll, which absorbs red light will re-emit red light (*Figure 1.6*). Invariably, however, the emitted light is of slightly higher wavelength (lower energy) than the light absorbed. This is because, within the fluorescence lifetime, there is time for vibrational and rotational loss of energy within each of the two energy states (see *Figure 1.5*). For example, a molecule excited to one of the upper vibrational levels within the excited state very rapidly ($< 10^{-12}$ sec) loses energy as heat to reach the lowest vibrational energy level in this state, and only subsequently (10^{-9} sec) remits a photon. Furthermore, during remission of a photon the excited state can revert to one of the upper vibrational levels of the ground (electronic) state, whence further energy is rapidly lost as heat (*Figure 1.10*). In summary, some of the absorbed energy will always be dissipated as heat; it is the remainder which emerges as the photon.

1.7 Fluorescence spectra and the quantitation of fluorescence

The fluorescence of a compound will vary as the wavelength of the excitation light is varied. For example, illumination at a strongly absorbed wavelength is likely to induce fluorescence emission while illumination at a wavelength that is poorly absorbed leads to little emission. Thus if we measure fluorescence at a given wavelength while varying the wavelength of the exciting light, we observe a wavelength dependence that reflects absorption of light by the compound, and this approximates to an absorption spectrum. This spectrum is called an excitation spectrum of the compound. (Note, however, it is fluorescence emission that is being measured.)

Alternatively, we may fix the wavelength of excitation (normally close to an absorption maximum) and observe how fluorescence emission varies with wavelength. This will generate a second spectrum, in this case called an emission spectrum. This will be different from the absorption spectrum as it reflects the energy of the downward transition of the excited state – which, as we have seen, yields slightly less energetic (longer wavelength) photons than those absorbed in the upward transition. Examples of an excitation and emission spectrum for NADH are shown in *Figure 1.11*.

By analogy with absorption, it should be possible to quantitate fluorescence by comparing the intensity of emitted light with the amount of light absorbed by the sample. This is expressed formally as the quantum yield, Q, where

Q = number of photons emitted/number of photons absorbed

However, this is a parameter rarely measured in biochemistry. This is largely because light is emitted from the sample in all directions, whereas measurements are made in a single direction and most emitted light is not detected (see *Figure 1.6*). In other words, the amount of emission measured will depend strongly on the geometry of the measuring system.

For most purposes, it is sufficient to express fluorescence emission in arbitrary units, which relate specifically to the instrumental set-up used. Thus, although fluorescence emission, like absorbance, is proportional to concentration (in dilute samples):

$$F = kc$$

the proportionality constant k will depend on the instrument as well as the compound under study. It is hence impossible to construct tables of 'fluorescence coefficients' which, like tables of absorptivities (*Table 1.1*) are portable from instrument to instrument. To calculate c from a measurement of fluorescence, it is always necessary to calibrate the instrument (i.e. find k for a particular instrumental set-up) using standard solutions of the compound under study (Chapter 8).

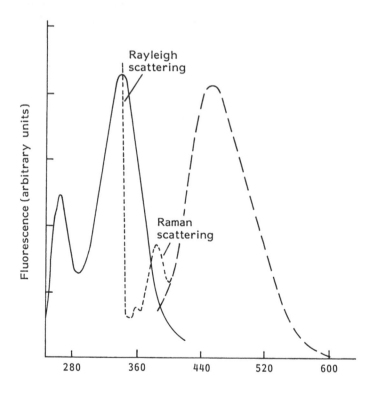

FIGURE 1.11:   *Fluorescence spectra of NADH. (——) Excitation wavelength varied, emission at 460 nm. (---) Emission wavelength varied, excitation at 340 nm. (···) Rayleigh and Raman scattering artifacts on measured emission spectrum (see Section 1.8).*

TABLE 1.1:   *Absorption properties of naturally occurring chromophores*

Chromophore	λ_{max} (nm)	ϵ at λ_{max} (M^{-1} cm^{-1} x 10^{-3})
Adenosine	260	15.4
NAD$^+$	260	18
Cytosine	267	6.1
Tyrosine	274	1.4
Guanosine	276	9.0
Tryptophan	280	5.6
NADH	340, 259	6.2, 14.4
Retinal	381	43.5
Riboflavin	445, 373	12.5, 10.6
β-carotene	451	142
Vitamin B12 (cyanocobalamin)	550, 361	8.7, 28.1
Heme a	598, 410	9.7, 69
Chlorophyll a	662, 430	91.2, 121

Only major λ_{max} values above 250 nm (near UV and visible) are given. Note that the absorption peaks may shift with the environment of the chromophore in biological systems (see Chapter 2). Because of the preponderance of chromophores which absorb at lower wavelengths, biochemical chromophores are normally characterized by their highest λ_{max}.

1.8 Light scattering revisited

We have seen that light impinging on a solution may be scattered, especially if the solution contains particles that are large relative to its wavelength. Light will thus emerge from the sample in a direction different from that of the incident beam (*Figure 1.2*).

If we measure light emerging from the solution along the incident beam, we find that its intensity is decreased ($I < I_o$). This is not due to the absorbance of the solution (as defined above); instead this effect is termed its turbidity. Turbidity differs from absorption in that it is dependent on particle size rather than on the chemical nature of the species, and also that it shows only a slight dependence on the wavelength of the incident light (*Figure 1.12*). Measuring absorbance in turbid samples (such as suspensions of biological membranes) requires special techniques as outlined in Chapter 4. Wherever possible, however, we try to measure absorption in optically clear solutions, where the loss in beam intensity due to light scattering is very small.

If we measure light emerging from a sample in a direction other than that of the incident beam, we can observe some of this scattered light. Again, this is not fluorescence emission, but simply diversion of the

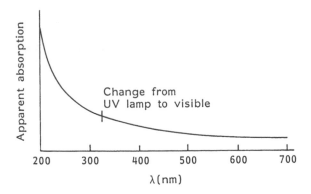

FIGURE 1.12: *Light scattering as a function of wavelength for liposome (200 nm diameter) suspension. Data provided by S. Benz.*

incoming beam. Most of this light has the same wavelength as the incoming light; this is due to Rayleigh scattering (elastic scattering), where the electron clouds of the molecules have oscillated with the frequency of the incident radiation. Since fluorescence emission occurs at a higher wavelength than the incident light, fluorescence emission can be distinguished from Rayleigh scattered light. In conventional fluorescence spectroscopy, Rayleigh scattering is minimized (but can never be completely abolished) by careful clarification of the solution.

A small fraction of the scattered light, however, emerges at a slightly different wavelength to the incident light; this is Raman scattering (inelastic scattering). This occurs when a molecule changes its vibrational state while the scattering is occurring. This light may interfere with measurements of fluorescence, as some Raman scattered light – like fluorescence emission – will emerge at a higher wavelength than the incident light. However, this event is rare at the molecular level; in dilute aqueous solutions, the only significant source of Raman scattered light will be the solvent, water. As the vibrational modes of water are known, we can predict that, in aqueous solutions, a band of Raman scattered light will be observed at a wavelength a fixed amount higher than that of the incident light. This can then be taken into account when fluorescence measurements are made.

Figure 1.11 indicates a spectrum of light emerging at right angles to the incident beam for the fluorescent molecule, NADH, excited at 340 nm. The Rayleigh and Raman bands due to scattered light are shown, together with the true fluorescence peak at 460 nm.

While light scattering is, in general, a liability in spectrophotometry and fluorometry, it can none the less be utilized for biochemical measurements. Because scattering increases with the number of particles in solution, turbidity is widely monitored, for example, as a convenient indication of bacterial growth in culture. More subtly, because it is sensitive to molecular size (in the 0.5–10 μm range), changes in light scattering can be used to monitor changes in the size of molecular aggregates. Thus measurements of turbidity (in a spectrophotometer) or of light scattered (in a fluorometer) have been used to monitor substrate transport by mitochondria (which swell osmotically as they take up solutes) and actin–myosin interactions (to form larger, actomyosin, complexes) in studies of muscle components.

2 What to Look at

2.1 Chromophores

Few biological compounds absorb light in the visible range. When they do, this property is often related to their function – either in trapping light energy to serve as a signal or as an energy source (e.g. rhodopsin, phytochrome, chlorophyll), or in pigmentation (e.g. melanin). An exception, however, occurs in the case of the transition metal complexes such as hemoglobin and the cytochromes (red), plastocyanin (blue), and so on. In this case, their color is irrelevant to their function, and represents an intrinsic property of the transition metal ion.

Rather more biochemical compounds absorb light in the near ultraviolet range. These include DNA, RNA and proteins, and smaller molecules like ATP and nicotinamide and flavin nucleotides. Here again absorption of light is unrelated to function. Indeed, absorption may even lead to damage and mutation *in vivo* (if, for example, DNA is irradiated with ultraviolet light).

Typically, absorption of light by these compounds is due to a relatively small grouping of atoms, termed a chromophore. For example, retinal is the chromophore in rhodopsin, heme in hemoglobin, the aromatic amino acids in proteins and the purine and pyrimidine bases in the nucleic acids and ATP. The structures of some biological chromophores are given in *Figure 2.1*. These groupings are chromophoric by virtue of either a conjugated double bond system and/or the presence of a transition metal.

Absorption of light occurs when the upper and lower electronic energy levels in a molecule are separated by a suitable energy gap (see Chapter 1). Saturated organic molecules, such as the sugar glucose, or the amino acid alanine, do not absorb light between 200 and 1000 nm.

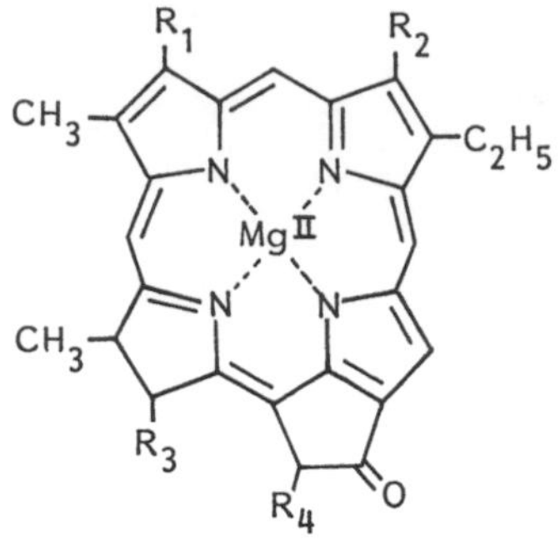

Chlorophyll

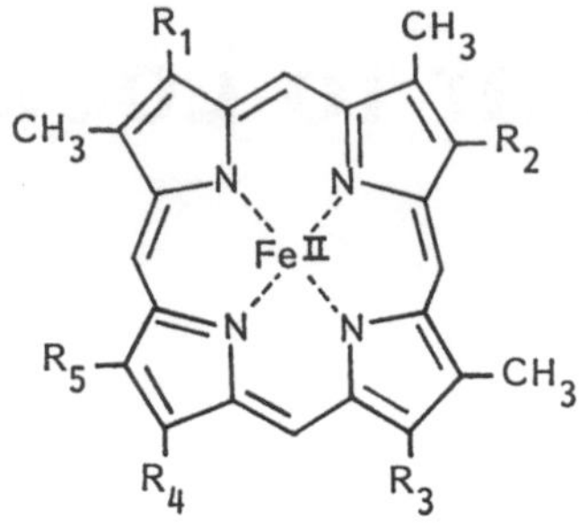

Heme

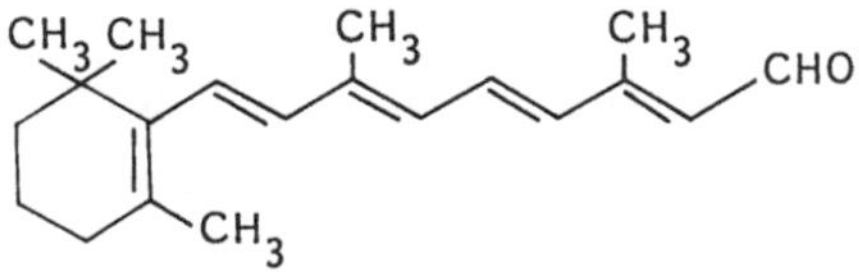

Retinal (all trans)

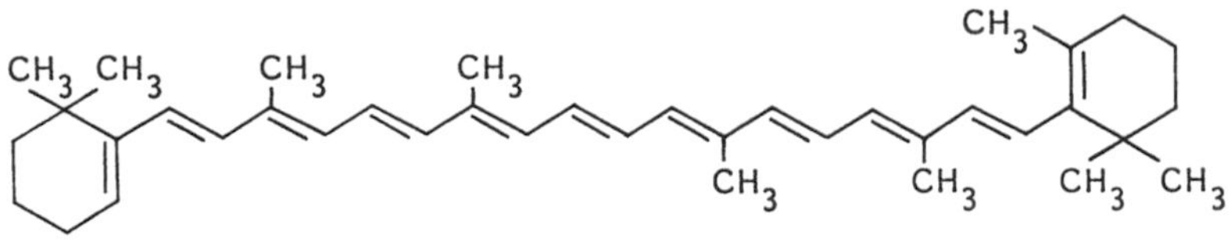

β-carotene

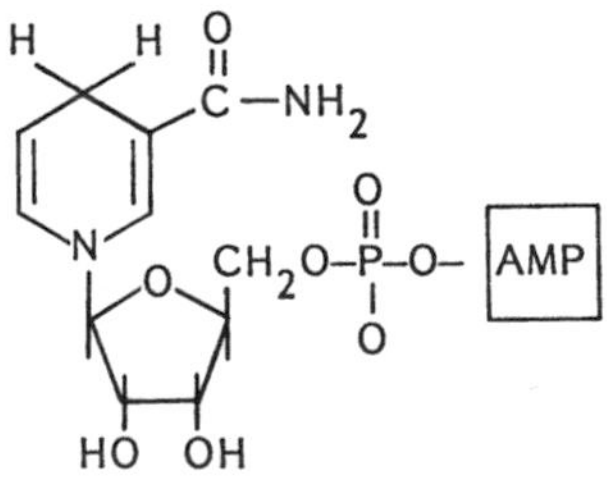

Nicotinamide adenine
dinucleotide
(reduced form)

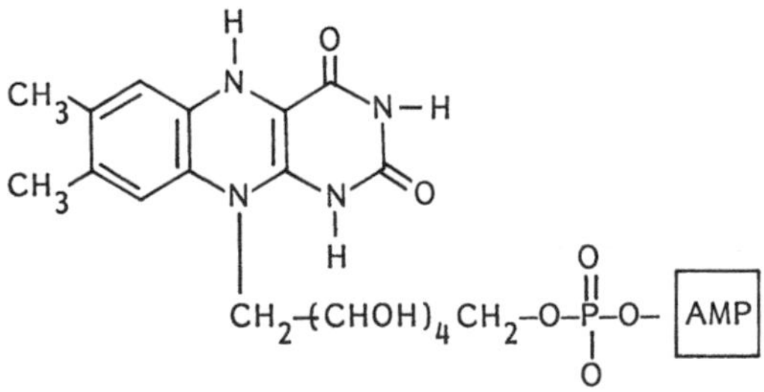

Flavin adenine dinucleotide
(reduced form)

FIGURE 2.1: *Biological chromophores. Structures of chlorophyll and heme have been simplified (side chains R_1, R_2...) to encompass a family of related structures.*

This is because the ground state and excited states are separated by too large an energy gap. In conjugated double bond systems, however, the energy levels of the delocalized π orbitals are close enough to those of the (empty) π^* orbitals to allow visible, or near ultraviolet light to effect a transition. In transition metal systems, transitions between the d orbitals can be induced (*Figure 2.2*).

In spectrophotometry, we derive information about molecules by observing the absorption of light by their chromophores. The importance of chromophores to the study of biological systems is emphasized by the discovery of one class of mitochondrial electron transfer proteins, the cytochromes, in 1888. These contain heme chromophores and were characterized and studied widely in the first half of the 20th century. However, it was not until 1956 that the first of a second class of electron transfer proteins, of similar importance and distribution, were discovered – the iron–sulfur proteins. These had remained unnoticed because (despite their iron content) they contained no convenient chromophore, giving only a very limited absorption of light in the visible region.

2.2 Fluorophores

We have seen, in Chapter 1, that a compound is fluorescent when (a) its molecules can absorb light (i.e. they contain chromophores) and also (b) this absorbed energy is re-emitted as photons, rather than (as is more usual) dissipated in vibrations and rotations as heat. By analogy with chromophores, the group of atoms in a molecule

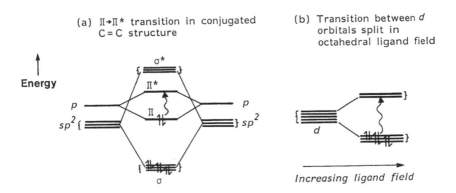

FIGURE 2.2: *Electronic transitions able to be driven by visible light.*

responsible for fluorescence emission is termed a fluorophore. Clearly, therefore, fluorophores comprise a subset of chromophores; they are thus even less common in biological systems.

Important naturally occurring fluorophores are the co-enzymes NADH and flavin (*Figure 2.1*), and the amino acid tryptophan (see Chapter 9). Chlorophyll too is naturally fluorescent in solution; in the cell however, the absorbed energy is trapped before re-emission can occur, and is used to drive photochemical processes. Its fluorescence is said to be quenched *in vivo*. Transition metals also quench fluorescence; the porphyrin ring itself shows fluorescence but when Fe^{II} is added – as in heme – fluorescence is abolished.

Like chromophores, fluorophores possess conjugated double bond systems (but not transition metals!). They also have to meet additional structural requirements. In particular, they must be motionally restricted, with few possibilities for rotations, vibrations etc., in order to limit energy loss as heat. This explains the rigid ring structures of the fluorophores in *Figure 2.1*. As another example, we can consider ATP; this has a large rigid ring system, but does not show fluorescence because energy can be dissipated via rotations of the amino group at C6 (*Figure 2.3a*). In the chemically modified derivative, etheno-ATP (ε-ATP) (*Figure 2.3b*), these rotations are abolished because the amino group is fixed in a ring system; in this case, motion is restricted and the molecule is strongly fluorescent.

2.3 Effect of environment on absorption

Measuring absorption and fluorescence spectra allows the experimenter to identify or characterize compounds under study. Measuring the amount of light absorbed or emitted allows an assessment of how much of the compound is present. Both of these approaches are widely used in the biological and clinical sciences. However, a third feature allows such measurements to yield information about the environment of the molecules under observation, that is, absorbing or fluorescing compounds can be used as probes of changes occurring in their vicinity.

Basically, any changes that alter the energy difference between the ground state and the excited state of a molecule will alter the wavelength of its absorbance or fluorescence. Since excitation involves a redistribution of electrons, a major factor affecting this energy difference is polarity of the environment. For example, when

(a)

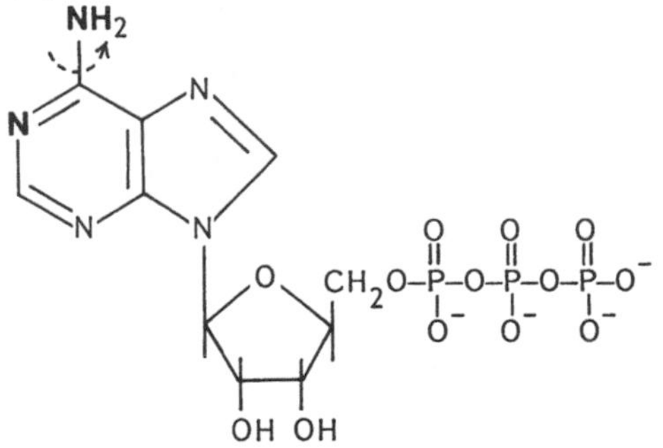

ATP (nonfluorescent)

(b)

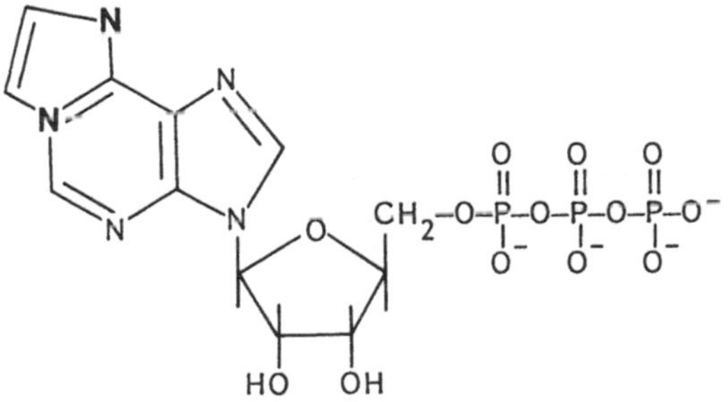

N_1–N_6 etheno–ATP (fluorescent)

FIGURE 2.3:   *Fluorescence properties of the adenine ring.*

NADH binds to lactate dehydrogenase (the nicotinamide chromophore moving from a polar, aqueous solvent to a more hydrophobic protein binding site), its absorption maximum shifts towards a lower (blue) wavelength (*Figure 2.4a*). The absorption shift can thus be used to monitor NADH binding to this enzyme.

The shift occurs because the excited state has a higher dipole than is present in the ground state. (This is frequently the case with organic molecules, due to the respective shapes of their π and π* orbitals.) In a less polar environment, therefore, the excited state is less stable, and more energetic photons are required to raise the molecule to this level; light of lower wavelength is absorbed. This effect is normally small (1–5 nm) as the shift in energy levels is relatively small (1–2 kJ mol^{-1}).

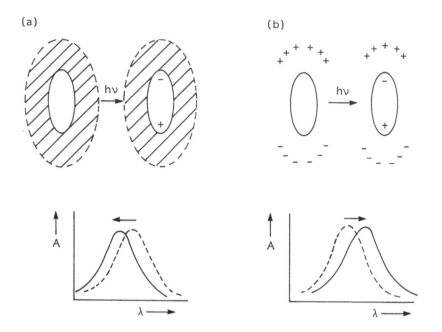

FIGURE 2.4: *(a) Ground state in apolar environment, excited state destabilized. Blue shift (to higher energy). (b) Ground state in electric field, excited state stabilized. Red shift (to lower energy).*

Similar effects are seen if the molecules are placed in an applied electric field. In this case the polar excited state is stabilized (*Figure 2.4b*), and absorption moves towards the red. This effect of electric field on absorption wavelength is known as the Stark effect, and can cause shifts of several nanometers at fields greater than 10^5 V cm^{-1}. Although fields of this magnitude may seem, at first sight, far removed from biological conditions, they are not; a typical biological membrane (5 nm = 5×10^{-7} cm across) with a typical transmembrane potential (100 mV) is experiencing fields of $(0.1 \text{ V}/5 \times 10^{-7} \text{ cm}) = 2 \times 10^5$ V cm^{-1}.

Hence, studies of potential-induced spectral shifts have turned out to be very useful in monitoring electrical potentials across biological membranes. For example, the carotenoids (similar in structure to β carotene in *Figure 2.1*) – red pigments found in the photosynthetic membranes of bacteria and chloroplasts – absorb green light at about 550 nm. Photosynthetic electron flow builds up an electrical gradient across these membranes. As outlined above, the Stark effect then shifts the absorption peak of the carotenoids towards higher wavelengths, and this red shift (the carotenoid shift) can be used to probe the development of a membrane potential (see Chapter 9).

2.4 Effect of environment on fluorescence

Like absorption wavelength, the emission wavelength of fluorophores is sensitive to their environment. Again, because of the (usually) increased polarity of the excited state, decreased polarity in the environment decreases its stability (*Figure 2.4a*) and shifts fluorescent emission to lower (blue) wavelengths. However, the effects of polarity on fluorescence are typically larger (10–20 nm shifts) than its effects on absorption (1–2 nm shifts). Thus fluorescence is more sensitive to environment than is absorption. This is because of the role of the solvent in stabilizing the electronic excited state. Absorption of a photon takes about 10^{-15} sec. Molecules do not move significantly in this period and distribution of solvent around the molecule after excitation is much as it was before excitation. Fluorescence emission, however, takes 10^{-9} sec, which, since molecules take only about 10^{-12} sec to rotate, allows ample time for the solvent to realign itself around the excited state and stabilize it (*Figure 2.5*). However, if the solvent is less polar, it will stabilize the induced dipoles in the excited state less well. Less energy is thus lost

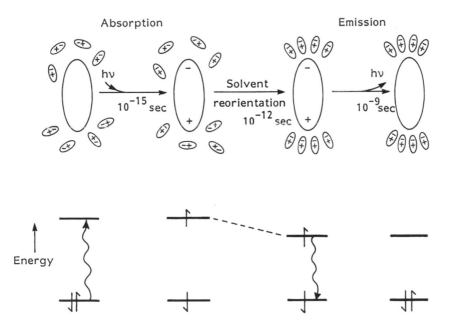

FIGURE 2.5: *Solvent reorientation in stabilizing the excited state of a molecule.*

as heat from the excited state, and more is released with the emitted photon – fluorescence emission shifts to lower wavelengths. To summarize, there is no time for solvent reorientation to occur during the absorption process, and thus absorption wavelength is less sensitive to environment than fluorescence emission.

A second effect of environment is on the quantum yield of fluorescence, that is, the intensity of fluorescence emission. (It is rare for the equivalent parameter for absorption – the molar absorptivity, ε_λ – to be much affected by the environment of the chromophore. A notable exception, however, is the nucleotide bases in DNA; when these are stacked in a double helical array, the delocalized π orbitals interact electronically and the absorptivity of the bases is some 40% less than would be exhibited in free solution. This is termed hypochromism.)

Various environmental factors may greatly increase or decrease the level of fluorescence at any given wavelength. This is because light emission and nonradiative decay of the excited state are competing processes (*Figure 2.6*) and if, for example, we decrease nonradiative decay, fluorescence will increase. Since a significant cause of nonradiative decay is collision of the excited fluorophore with solvent molecules, shielding the fluorophore from solvent will commonly result in an increase in fluorescence emission. The effect of the environment on NADH fluorescence when NADH binds to lactate dehydrogenase is shown in *Figure 2.7a*; both a blue shift of the emission band and an increase in its intensity, is observed. The magnitude of these changes is again much greater than the change in absorption spectrum on binding (not shown).

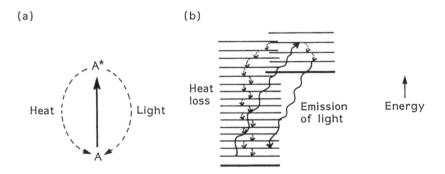

(a) (b)

FIGURE 2.6: *Competition between light emission and heat loss (nonradiative decay) for energy of the excited state. (a) Schematic transitions between ground state A and excited state A*. (b) Representation of energy levels involved (thick lines indicate electronic energy levels; thin lines represent vibrational and rotational energy levels).*

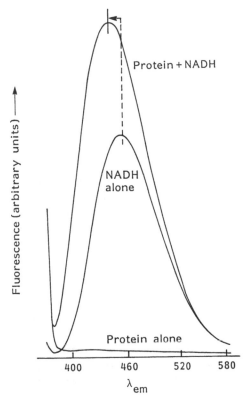

FIGURE 2.7:   *Fluorescence emission spectra of NADH ± lactate dehydrogenase. Note the enhancement of fluorescence of NADH, and the blue shift of the emission peak on binding NADH to the enzyme.*

Alternatively, it is possible to add to a solution of the fluorophore, compounds which promote nonradiative decay; these work by being more efficient than water at dissipating the energy of the excited state. Examples of such compounds, known as quenchers of fluorescence, are iodide ion (I⁻) and acrylamide. Such compounds can be used to probe exposure of fluorophores (e.g. tryptophan in protein molecules) to the solvent.

It is also possible that the energy that would have been emitted as fluorescence is transferred through space to another group. For this to occur: (i) the recipient group itself (acceptor) must be a chromophore, with an absorption band overlapping with the emission band of the donor, and (ii) the acceptor must be close to (within 5 nm of) and aligned with the donor. Transfer is most easily observed if the acceptor is itself a fluorophore; in this case, excitation at wavelengths suitable for the donor will lead to an emission spectrum characteristic of the acceptor molecule (*Figure 2.8*). From our knowledge of fluorescence emission, we can predict that this emission band must be even further towards the red (compared with the excitation wavelength) than that of the donor.

This energy transfer occurs without re-emission of the photon, and is termed 'resonance energy transfer' or Förster energy transfer. Instead, it involves direct interaction between the two molecular dipoles during their electronic transitions. This means that the transfer is highly dependent on distance in proportion to $1/R^6$ (where R is the distance of separation) – hence its short range (above). However, the distances involved (2–5 nm) are within the range of sizes for biological macromolecules, and this has led to the use of resonance energy transfer measurements as a type of 'molecular ruler', to estimate distances within DNA proteins and membranes (see Chapter 9).

2.5 Sensitivity and detection limits

We have seen that fluorescence emission changes more dramatically with environment, in both wavelength and magnitude, than does absorption. This means that, if we wish to probe changes in a

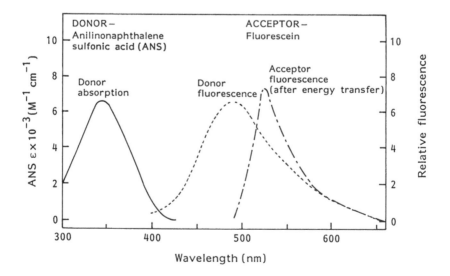

FIGURE 2.8: *Shift of emission wavelength after energy transfer between two fluorophores. Redrawn from Hirs and Tunasheff (eds) (1979)* Methods Enzymol., **48, 362,** *with permission from Academic Press.*

molecule's environment, fluorescence measurements are likely to yield clearer information than absorption measurements. In contrast, if we wish to assess accurately concentrations in solution, or follow chemical changes, the method of choice would be absorption measurement, where the relevant parameters (ε_λ, λ_{max}) are more stable and vary little with different conditions of measurement.

However, another factor limiting our choice is the absolute sensitivity (detection limit) of each method compared with the amount of material available. In principle, fluorescence measurements can detect much smaller amounts of material than can absorption measurements (10^{-15} mol as opposed to 10^{-9} mol) due to the level of the background light measured in each case.

This principle is illustrated in *Figure 2.9* and can be explained as follows. The factor that limits detection of any change is the fractional difference this change produces in the background level of the signal. When measuring absorption, we are measuring a decrease in light intensity from an initial high value (I_o); thus absorption of 10^4 photons in every 10^6, for example, would produce a barely detectable 1% change in transmission (absorbance, $A = 0.004$). When measuring fluorescence, on the contrary, these same 10^4 photons, if emitted, would be measured against a background of zero photons – an infinite percentage increase. This means that, even when measuring concentrations, fluorescence measurements must be used if very low concentrations are involved. Note, however, that because of the larger effects of environment on fluorescence – in particular, the effect of temperature – fluorescence measurements of concentration require much more careful control of experimental conditions than do absorption measurements.

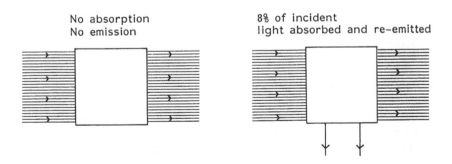

No absorption
No emission

8% of incident
light absorbed and re-emitted

FIGURE 2.9: *Sensitivity of detection of absorption and fluorescence. Absorption of 8% of incident rays is difficult to see in this diagram, but emission of this light is easily seen.*

2.6 Intrinsic and extrinsic probes

Naturally occurring chromophores, such as those depicted in *Figure 2.1* are very useful for following events within biological systems, such as the reduction of NAD+ or the oxygenation of hemoglobin. However, naturally occurring (or intrinsic) chromophores are rare; most biological molecules do not absorb light in the region 200–1000 nm, and very few indeed are fluorescent. When no intrinsic chromophore is present however, it may be possible to utilize the power of a spectrophotometric method by introducing a novel chromophore – an extrinsic probe – into the system. Providing this abiological molecule does not perturb the system under study, measurement of its optical properties and their variations can give useful information about changes in the system under study.

Various types of chromophore are discussed below. The list is not meant to be exhaustive, but rather to give some idea of the types of approach available. Other examples will appear in subsequent chapters.

2.6.1 Probes for labeling proteins

Various compounds, like fluorescein and rhodamine, are highly fluorescent but bear little relationship to biological molecules. They can be linked chemically to proteins by attaching to them a reactive chemical group which will then react with amino acid side chains of proteins. The resultant fluorescent protein can be used simply as an aid to increase the sensitivity of detection (for example, the use of fluorescein-labeled antibodies) or, by utilizing the environmental sensitivity of the probes, to investigate conformational changes in proteins or other macromolecular assemblies. Examples of suitable reagents are shown in *Figure 2.10a*. Note the large ring structures of the fluorophore in each case.

2.6.2 Ligand analogs

We have seen how the absorption and fluorescence of NADH change as it binds to lactate dehydrogenase. These changes can be used to measure binding constants and otherwise probe the nature of the binding site for the ligand under study.

While most biological ligands are not chromophoric, it is often possible, by judicious chemical modification, to change them into

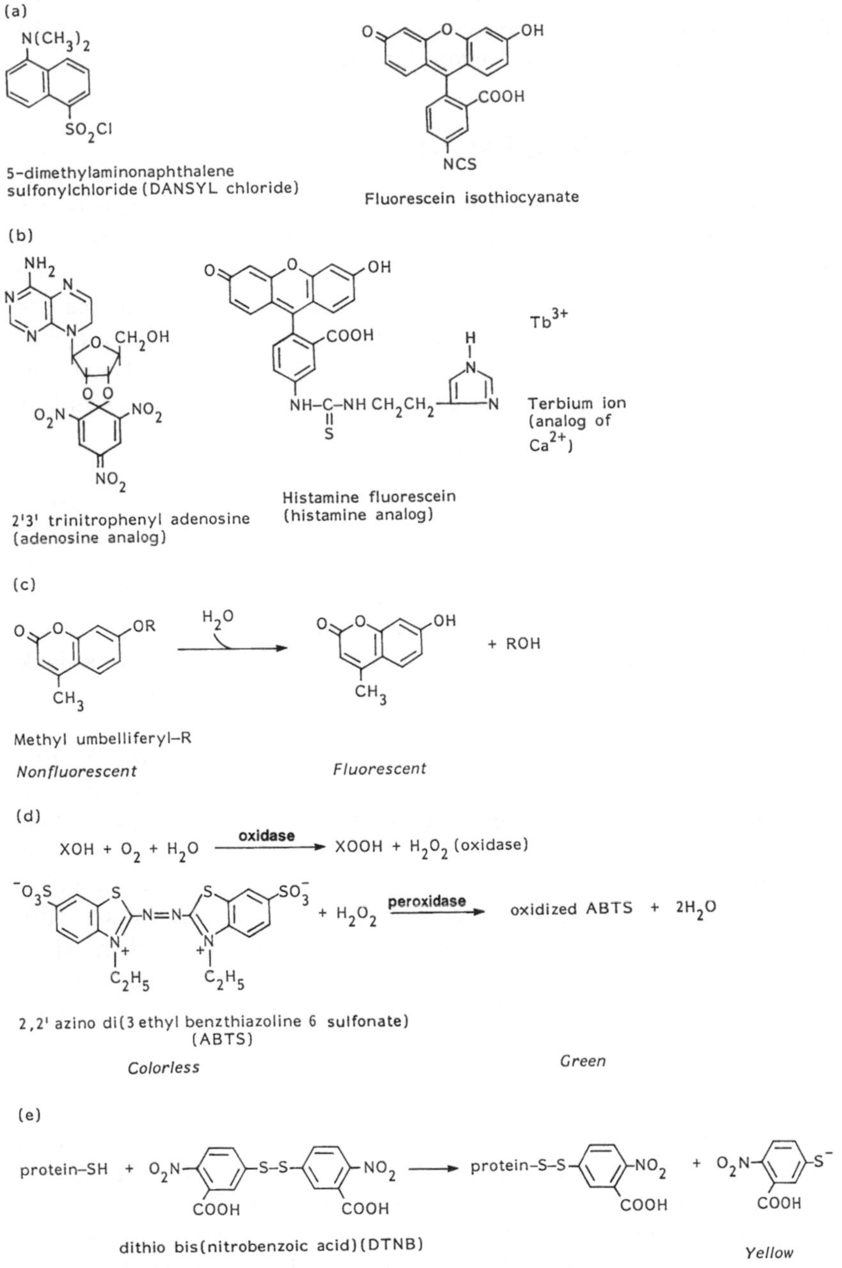

(a)

5-dimethylaminonaphthalene
sulfonylchloride (DANSYL chloride)

NCS

Fluorescein isothiocyanate

(b)

Tb^{3+}

Terbium ion
(analog of
Ca^{2+})

2'3' trinitrophenyl adenosine
(adenosine analog)

Histamine fluorescein
(histamine analog)

(c)

Methyl umbelliferyl–R

Nonfluorescent

Fluorescent

+ ROH

(d)

$$XOH + O_2 + H_2O \xrightarrow{\text{oxidase}} XOOH + H_2O_2 \text{ (oxidase)}$$

$$+ H_2O_2 \xrightarrow{\text{peroxidase}} \text{oxidized ABTS} + 2H_2O$$

2,2' azino di(3 ethyl benzthiazoline 6 sulfonate)
(ABTS)

Colorless

Green

(e)

$$\text{protein–SH} + O_2N-\bigcirc-S-S-\bigcirc-NO_2 \longrightarrow \text{protein–S–S}-\bigcirc-NO_2 + O_2N-\bigcirc-S^-$$

dithio bis(nitrobenzoic acid)(DTNB)

Yellow

FIGURE 2.10: *Examples of extrinsic chromophores for biochemical studies (see also* Table 2.1). *(a) For covalent labeling of proteins; (b) for occupying ligand binding sites on proteins; (c) for assaying hydrolases; (d) for assaying oxidases; (e) for quantitating cysteine residues in proteins.*

colored or fluorescent compounds without losing their biological activity (*Figure 2.10b*). These extrinsic probes can then be used to study the binding site of the normal ligand.

2.6.3 Chromogenic substrates

Enzyme activities are followed by measuring the increase in product (or decrease in substrate) concentration with time. If the substrate or product, but not both, is colored (or fluorescent), activity can be conveniently followed by monitoring the change in absorbance (or fluorescence) of the solution. Dehydrogenase activity, for example, can be followed by measuring the conversion of NAD^+ (no absorption above 300 nm) to NADH (strong absorption at 340 nm; fluorescent at 460 nm) (see Chapter 8).

If, as is more common, the natural substrate or product is neither colored nor fluorescent, one might be converted into a conveniently colored compound. For example, in measuring the hydrolysis of phosphate esters, the production of phosphate can be followed by stopping the reaction and complexing any free phosphate with molybdate ions. In strong acid, this complex can be reduced to a dark blue compound that can be measured spectrophotometrically.

A more convenient approach is to design an artificial, chromogenic, substrate which itself changes color on conversion. A large number of such substrates have been designed for hydrolytic enzymes, based on esters of nitrophenol (which gives a yellow product) or methylumbelliferone (7-hydroxy, 4-methyl coumarin – which gives a fluorescent product). In both these cases, the anion is strongly chromophoric, with the negative change delocalizing around the ring, while the ester shows little coloration (*Figure 2.10c*). The acyl moiety of the substrate can be varied to match the specificity of the enzyme under study.

Whether it is better to choose a fluorescent or simply a colored reagent depends on the activity of the enzyme to be measured. The increased sensitivity of fluorescence measurements means that the umbelliferyl derivative would be the substrate of choice for low enzyme activities. Conversely, for higher activities, the insensitivity of an absorbance-based assay to interference by environmental factors would be the deciding factor. *Table 2.1* indicates the types of enzyme for which chromogenic substrates are commercially available.

In a related technique, numbers of molecules, or reactive groups within macromolecules, can be assessed quantitively if a convenient chromogenic reagent (i.e. one which gives a colored compound after reaction with the molecules) is available. For monitoring the number of sulfhydryl groups in a protein, for example, a convenient compound is dithio-*bis*-nitrobenzoic acid (DTNB). This reacts with sulfhydryl groups yielding a nitrothiobenzoate anion, which, like nitrophenol, is yellow as the anion (*Figure 2.10d*). Other examples include trinitro-benzene sulfonate (TNBS) and diethyl pyrocarbonate (DEPC) for monitoring amino and imidazole groups, respectively, within proteins.

2.7 Indicator molecules

2.7.1 Indicators in free solution

Classical pH indicators provide a colorimetric indication of the pH of a solution. Thus, in biological systems where a pH change is to be observed, such an indicator can be used. This allows observations with very high time resolution, less than 1 μsec, as opposed to 1–10 sec, using a conventional pH electrode.

TABLE 2.1: *Chromogenic substrates for enzyme reactions*

Chromophore produced	Enzymes	Examples	Detection
o or *p* Nitrophenolate	Esterases	*o*-Nitrophenyl β-galactoside (for β-galactosidase) *p*-Nitrophenyl phosphate (for alkaline phosphatase)	A_{405}
4-Methyl umbelliferin (7-hydroxy, 4-methyl coumarin)	Esterases	7-Acetoxy, 4-methyl coumarin (for acetylcholinesterase) 4-Methyl umbelliferyl sulfate (for aryl sulfatase)	F_{444}
Resorufin	Esterases	1,2-O-dilauroyl-glyceryl 3-glutaric acid resorufin ester (for lipases)	A_{572}, F_{583}
Hippurate (benzoyl-glycinate)	Peptidases	Hippuryl arginine (for carboxypeptidase B)	A_{254}
Nitroaniline	Proteases	Benzoylarginine 4-nitranilide (for trypsin)	A_{253}
7-Amino 4-methyl coumarin (AMC)	Proteases	Benzyloxycarbonyl-arg-arg-AMC (for cathepsin B)	F_{450}
Rhodamine-110 (R110)	Proteases	(Benzyloxycarbonyl-val-pro-arg)$_2$ - R110 (for thrombin)	F_{572}
Oxidized ABTS®	Peroxidases	2,2'-azino-di-[3'-ethylbenz thiazoline sulfonate] (ABTS®)	A_{405}
Fluorescein	Peroxidases	2',7' dichloro*dihydro*fluorescein (for horseradish peroxidase)	F_{529}

Examples of various classes of compound are given. The list is not intended to be exhaustive – a wide variety of esters and peptides are commercially available.

Phenol red has been widely used in this role. It has a convenient pK of ≈7 so it can respond to pH changes within the physiological pH range, and it binds poorly to proteins so it is unlikely to perturb the biochemical system under study. An example of its use includes measuring the uptake of H⁺ ions into vesicles, such as the thylakoid vesicles of chloroplasts under illumination. H⁺ uptake causes a rise in pH outside the vesicles in the vicinity of the indicator, and a consequent color change. Phenol red has also been used to monitor ligand binding to protons – ATP binding to myosin, for example, results in the rapid release of H⁺ into solution which can be detected by this indicator.

Alternatively, movement of H⁺ into, or out of, vesicles can be monitored by using a distribution probe. 9-Aminoacridine (9-AA; *Figure 2.11c*), for example, is a weak base that can cross biological membranes. When protonated, however, it is impermeant and hence it accumulates on the low pH side of membranes (e.g. inside illuminated chloroplast membranes) (*Figure 2.11a*). This is accompanied by a marked decrease (quenching) in fluorescence (*Figure 2.11b*). Thus the fluorescence of 9-AA (and related compounds) can be used to monitor H⁺ accumulation in vesicles (see Chapter 9).

2.7.2 Intracellular indicators

Over the past few years, there has been considerable interest in determining the concentration of ions free within cells, and even within individual cell compartments. Breaking open the cells to have access to the contents is not an option in these cases. Firstly, it is likely to disrupt the intracellular ion distribution between compartments (especially in the case of H⁺ ions). Secondly, chemical analysis of the contents reveals the total ion concentration, whereas (in the case of Ca²⁺, for example) more than 90% of the ions are bound to protein or other molecules and do not contribute to the free concentration. Thirdly, breaking the cell makes it impossible to follow intracellular changes in ion distribution as they occur, during cellular functions such as muscle contraction. Thus attempts have been made to develop indicators (by analogy with pH indicators) which change color with changes in ion concentration and can signal such changes for spectrophotometric (or fluorometric) detection.

Development of suitable molecules has involved a great deal of sophisticated chemistry. Not only must the indicator change color with the concentration of the ion under study, it must have a suitable affinity for the ion (i.e. the amount of ion complexed to it must change

(a)

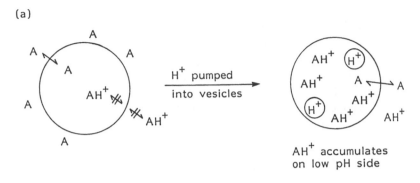

(b)

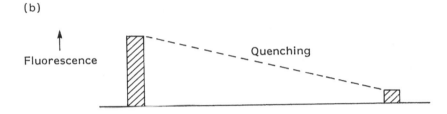

(c)

9−amino acridine

FIGURE 2.11: *H⁺ uptake into vesicles leads (a) to accumulation of AH⁺ inside (and decrease of total free A), and hence (b) decrease in fluorescence; (c) shows the structure of α aminoacridine.*

over a physiological range of concentrations). It must be able to enter the cell, and yet once inside the cell, it must not diffuse out again; this will allow it to measure changes in the cytoplasm without interference from ions in the extracellular milieu. It should be specific for the ion under study, so that it reports only relevant changes to the observer. It must be detectable at low concentrations so that the cell is hardly perturbed by its presence (hence normally a fluorescent molecule is used); and, of course, it must be nontoxic to the cell.

Perhaps the best known of such molecules is the Ca^{2+} indicator, fura-2 (*Figure 2.12*), which has revolutionized our understanding of the behavior of Ca^{2+} within cells. Typically of this class of molecules, it is introduced into the cells as an uncharged ester, which crosses the plasma membrane. Inside the cells, this is hydrolyzed (by nonspecific esterases) revealing carboxylate groups which can bind Ca^{2+}, simultaneously trapping the compound inside the cell (because charged compounds cannot traverse biological membranes). On binding Ca^{2+}, its excitation wavelength shifts, and thus changes in $[Ca^{2+}]$ can be monitored (see *Figure 4.10*).

Related indicators include indo-1 and fluo-3 for intracellular Ca^{2+}. All these compounds have pCa^{2+} values of ≈ 6, and thus will respond to $[Ca^{2+}]$ in the physiological range of 10^{-7}–10^{-5} M. Examples of intracellular pH indicators are ionizable fluorescein derivatives with pK_a values of $\approx$ 7, such as 2′,7′-*bis*-(2-carboxyethyl)-5-carboxyfluorescein (BCEC-fluorescein) and 10-amino 3-hydroxy-spiro [7H-benzo (c) xanthene-7,1′ (3H)-isobenzo furan]-3′-one (SNARF™), which change their fluorescence emission between the protonated and deprotonated states.

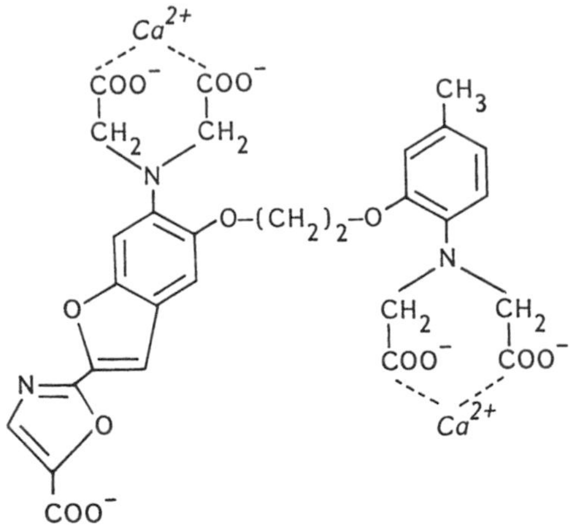

FIGURE 2.12: *1-[2-(5-carboxyoxazol-2-yl)-6-aminobenzofuran-5-oxy]-2-(2′-amino-5′-methylphenoxy)-ethane-N,N,N′,N′ tetraacetic acid (fura-2). The positions of Ca²⁺ binding are indicated. Only the Ca²⁺ complex is fluorescent.*

3   Spectrophotometer Design

3.1 Components

A basic spectrophotometer is shown diagrammatically in *Figure 3.1a*. Light from a source (typically a lamp) is made monochromatic and passed through the absorbing sample. The amount of light emerging is then measured using a photosensitive device (such as a photomultiplier). These components are organized along an optical bench. Transmitted light is measured along the direction of illumination.

A basic spectrofluorometer shares the same components, the difference lying in the geometry of the system. Here the emitted light is measured along a different direction from illumination – commonly at right angles to (*Figure 3.1b*) or above the plane of the illumination beam.

Additional components include collimator lenses or mirrors (for condensing or focusing the beam), devices for maintaining the sample under constant conditions (thermostats, stirrers, etc.) and instruments for recording the electrical signal produced.

3.2 Light sources

3.2.1 Continuous illumination

An ideal spectrophotometer lamp should produce a constant intensity of illumination over a wide range of wavelengths. (The level of

(a)

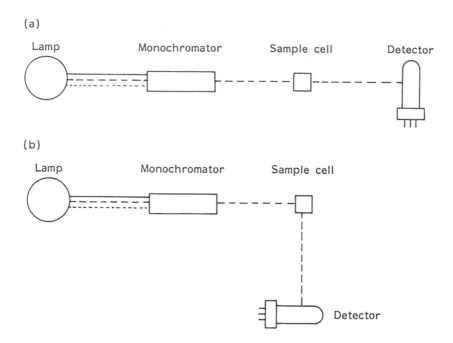

FIGURE 3.1: *(a) Basic absorption spectrophotometer. (b) Basic right angle spectrofluorometer.*

intensity is less important, because normally most of the light falling on the sample is measured at the detector.) The tungsten filament lamp, producing more or less black body radiation performs well in the visible and near infrared region (350–900 nm), although the tungsten halogen lamp (also known as the quartz–halogen lamp) has a higher intensity in the blue and is now preferred by many manufacturers.

Neither of these has sufficient intensity in the ultraviolet for measurements down to 200 nm; in this case, the deuterium arc lamp is almost universally employed (*Figure 3.2*). Most modern instruments contain both quartz–halogen and deuterium lamps, and switch (automatically?) between the two as the wavelength of measurement drops below 350 nm (see *Figure 1.12*).

Other vapor emission (arc) lamps, such as mercury and xenon lamps, produce more intense illumination, but most of the energy is concentrated into narrow bands of wavelength (emission lines). Mercury lamps are occasionally used for fixed wavelength absorption measurements at 254 nm, in particular when UV monitoring of a chromatographic eluate is required (for example, in high-performance

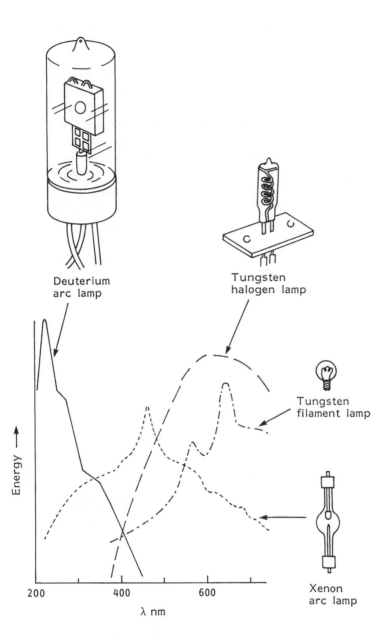

FIGURE 3.2:   *UV/visible light sources, and their emission characteristics.*

liquid chromatography, hplc), but are unsuitable when a range of wavelengths are required.

Fluorescence emission is proportional to the intensity of the incident light. Thus, in fluorometers, the brightness of a light source is of greater importance than wavelength uniformity. Thus, in contrast to spectrophotometers, vapor emission lamps are common in commercial fluorometers. Xenon arcs have a more continuous emission over a range of wavelengths and are therefore preferred over mercury arc lamps. However, the number of photons emitted still varies significantly with wavelength (*Figure 3.2*), and this needs to be considered when measuring the dependence of emission on absorption wavelength (emission spectra) (see Chapter 7, Section 7.9).

Lasers provide highly monochromatic light sources and as such are not suitable for general purpose spectrophotometers or fluorometers. However, their high intensity has led to their use in more specialized applications, such as fluorescence-activated cell sorting, or in other systems where very small numbers of molecules are to be detected. Pulsed lasers (10^{-15} sec pulse) are also used for following very rapid events; a typical lamp flash, from a xenon flash lamp, lasts 10^{-9} sec or longer.

3.2.2 Pulsed illumination

In many instruments, the light falling on the sample is modulated at a known frequency. This has the advantage that the detector receives light modulated at the same frequency, and allows the use of a.c. amplifiers into the electronic circuits. These have a better performance than d.c. amplifiers. This also allows the amplifier to 'recognize' light coming from the illumination lamp, while extraneous light (unlikely to be modulated at the same frequency) is ignored. A third advantage, particularly with intense light sources, is that the sample is illuminated for less time overall and chemical effects, such as photobleaching (see Chapter 4), are minimal.

With tungsten filament and deuterium arc lamps, pulsed illumination is achieved by keeping the lamp on continuously but modulating the beam by a rotating mechanical chopper (see Chapter 4, *Figure 4.5*) which interrupts the beam at intervals. Quartz–halogen and xenon lamps (in fluorometers) are commonly pulsed electronically. This has the additional advantage, to the operator, that the photochemical production of ozone is greatly reduced compared with a lamp giving continuous illumination.

3.3 Wavelength selection

Filters may be used to select a wavelength for illumination. Colored glass filters typically allow a broad band of wavelengths to pass through (50–100 nm wide). Interference filters, in contrast, can reduce this band width to 5–10 nm. A number of interference filters with wavelength bandwidths (bandpass) of 5–10 nm are commercially available from Oriel and, placed between the lamp and the sample, can provide a narrow band of illumination wavelengths (*Figure 3.3*). Filters are convenient in fixed wavelength applications (such as monitoring chromatographic eluates or in fluorescence microscopy), especially where space is limited. In addition, they are employed automatically within some instruments to eliminate unwanted radiation (including higher order harmonic reflections from a grating monochromators). However, even the simplest commercial spectrophotometers today provide a continuously variable wavelength of illumination which is impossible to achieve using individual filters.

Instead, this is accomplished with a monochromator. This is made up of two elements, one for dispersing the illuminating beam into a spectrum of separate wavelengths (a prism or grating) and one for selecting a narrow band of wavelengths from the spectrum (a 'slit') (*Figure 3.4*). For dispersion, modern instruments almost invariably use gratings, which utilize interference effects for dispersion, rather than prisms which utilize refraction effects. This is because

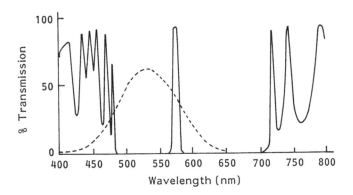

FIGURE 3.3: *Transmission characteristics of band pass filters: (- - -) interference filter (peak 576 nm, half height width 9nm); (···) colored glass filter (peak 530 nm, half height width ≈ 100 nm). Note narrow band width of the interference filter. Transmission at <500 nm and >700 nm can be removed by adding cut off filters to the coating.*

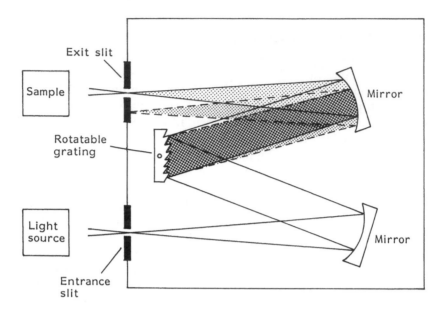

FIGURE 3.4: *Czerny-Turner monochromator, showing dispersion of white light at grating and selection of appropriate wavelength by exit slit.*

holographic (optically ruled) gratings can be produced more precisely and more cheaply than prisms. In addition, they are much lighter in weight, allowing them to be moved more quickly during wavelength scanning than is possible for a prism.

A slit is basically a hole in an optically opaque material, such as black anodized aluminum. Slits are typically 0.1–1 mm across with parallel vertical sides. The width of a slit, and its position relative to the grating, will set the width of the wavelength band passed (the 'spectral bandwidth'), which is typically 1–10 nm. (*Beware* – slit widths in manufacturers' handbooks may be given either as physical width or as spectral bandwidth.) A popular arrangement for a monochromator is the Czerny–Turner grating monochromator, which, as can be seen in *Figure 3.4*, incorporates both these elements.

Some instruments allow the width of a slit to be varied manually. This allows the operator to balance the opposing considerations of spectral bandwidth (minimized for high wavelength resolution) and intensity of illumination (maximized for high signal/noise ratio). With highly absorbing solutions, for example, it may be necessary to open the slit wide (and hence lose wavelength resolution) so that sufficient light can reach the detector for a measurable signal. In many commercial spectrophotometers, however, the slit width is varied automatically and is inaccessible to the operator.

3.4 The sample compartment

3.4.1 Measuring individual samples

The absorbing or fluorescent sample must be held precisely in the light beam, between the monochromator and the detector. For solution measurements, this requirement is satisfied by a combination of two elements, a cuvette, which holds the solution to be measured, and a cuvette holder, which rigidly positions the cuvette in the light beam.

The cuvette, or sample cell, is a vessel with two transparent sides for the entry and emergence of the light beam. Glass is transparent in the visible region (350–900 nm), but absorbs strongly in the ultraviolet; below 350 nm quartz or fused silica cuvettes can be used down to 190 nm. Alternatively, plastic (disposable) cuvettes are now widely available, made from polystyrene (suitable for wavelengths above 350 nm) or polyacrylate (for wavelengths above 250 nm). The transmission characteristics of these materials are shown in *Figure 3.5*. Cuvettes for absorption measurements typically have one pair of (opposite) sides transparent, while for fluorescence measurements (with right angled geometry), all four sides are clear.

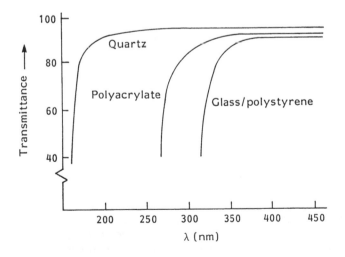

FIGURE 3.5: *Transmission characteristics of cuvette materials.*

The standard cuvette for biochemical work has dimensions 1 cm (light path) × 1 cm × 4 cm, hence comfortably holding a volume of 3 ml (*Figure 3.6a*). Modifications of this design allow smaller volumes to be used (*Figure 3.6b*) (down to 5 μl, for molecular biology work, in the case of the Beckman Capillary Ultramicrocell), or allow a solution to flow continuously through the cell (*Figure 3.6c*) (see below). Where small volumes are to be measured it is important to prevent light reaching the detector without passing through the sample [a component of stray light (see Chapter 4)]. This is achieved by (i) masking the sample compartment so that no other light path is

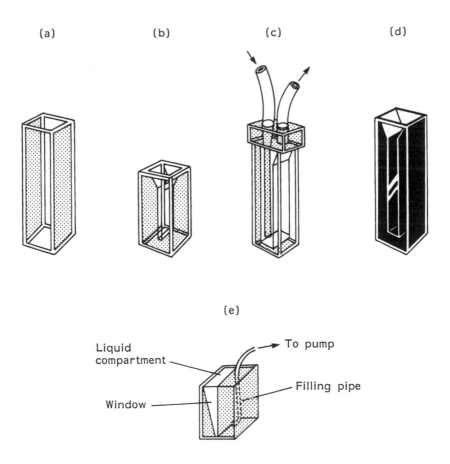

FIGURE 3.6: *Cuvettes for absorbance measurements. (a) Standard cuvette (internal vol. 3.5 ml); (b) micro cuvette (internal vol. 0.4 ml); (c) flow through cell (internal vol. 1.8 ml); (d) self masking, semi-micro cuvette; (e) 'sipper cell' cuvette, for intermittent filling/emptying. Note that all have base dimensions 1 cm × 1 cm to fit into standard cuvette holder.*

available (*Figure 3.6b*), and (ii) focusing the measuring beam precisely on to the sample (see Chapter 4).

The cuvette holder is essentially a metal frame, into which the cuvette fits snugly, which is rigidly positioned in the light path (e.g. by being bolted to the base of the sample compartment). Elaborations include holders for multiple (up to 12) cuvettes and thermostatted cuvette holders, in which the frame is hollow and (usually) can be perfused with water from a thermostatically controlled bath. A constant temperature is important when measuring reaction rates (kinetics), which are highly temperature dependent, and also when measuring static fluorescence, since fluorescence quenching is similarly sensitive to temperature. It is notable that glass or quartz cuvettes conduct heat much better than plastic cuvettes, and are preferred for applications where thermostatting is involved.

The cuvette and holder are positioned in a box with a light-tight lid, the sample compartment. Since, in modern instruments, the optical and electronic components are inaccessible, this is the region of the spectrophotometer/fluorometer which is most familiar to the user. It is also the region that is likely to come into contact with solutions etc., as samples are changed and spillages occur. Unfortunately, while considerable improvements in optics and detection have occurred over the years, the design of the sample compartment and its components has been given less attention. Inaccessibility of the sample holder, malfunctioning of its mechanical components or thermostatting and/or difficulties in cleaning it are still problems of many spectrophotometers today.

3.4.2 Measuring several samples

It is common in biochemistry to perform the same measurement on a number of samples – to locate, for example, where a particular substance elutes in a series of column fractions, or to measure the levels of a metabolite in blood samples from various individuals. This necessitates repeated emptying, washing and refilling if a single cuvette is used. Spectrophotometers may be fitted with various devices to handle multiple samples.

(i) A holder carrying several cuvettes, each of which can be separately introduced into the light beam (either manually or by a motor driven device). In absorption spectrophotometers, these multiple cell holders are typically linear and move along a linear metal beam to position each cuvette in turn in the measuring beam; in fluorometers (with right angle geometry) they rotate around a central axis.

(ii) A flow-through cell, in which the cuvette is fixed in the instrument and the samples are continuously pumped through the light path. This is commonly used for monitoring column eluants (*Figure 3.6c*), although the principle is adapted in the continuous flow instruments used for measuring multiple samples in clinical laboratories.

(iii) A sipper device. This is somewhat of a halfway house between the two. Here the cuvette (*Figure 3.6e*) is fixed, and rapidly filled by pumping the sample from a reaction vessel. After measurement, the liquid is pumped out to waste, and the next sample pumped in.

An alternative route to the facile measurement of multiple samples uses a vertical, rather than horizontal light path, and is described in Section 4.4.

3.5 Detectors

Light transmitted by, or emitted from, the sample falls on a detector, which in response produces an electrical signal which can be measured. Detectors are either photomultipliers or photodiodes.

A photomultiplier is an electronic valve in which light impinging on a photocathode leads to electrons being displaced from it. These are then accelerated, displacing further electrons from charged grids (dynodes), so that an amplified current arrives at the anode (*Figure 3.7*) (the device is strictly an 'electron multiplier', rather than a 'photomultiplier'). Amplification can be very high, and in some cases a single photon event can be detected as a pulse at the anode. For most applications, photomultipliers are set up so that the current produced is proportional to the intensity of light falling on the photocathode (an analog signal). In some cases, however, the circuitry may be configured to count pulses (= photons), that is, to produce a digital signal. This latter system is used typically where low light intensities are expected, for example in some fluorometers (e.g. PTI RF series), rather than in spectrophotometers.

Photomultipliers are highly sensitive, and are used widely in standard laboratory spectrophotometers and fluorometers. However, they are bulky, fragile and require a high voltage supply, making them unsuitable for small or portable instruments. In addition their operation depends on the 'photoelectric effect', where the energy of the impinging photon must be sufficient to displace an electron from the cathode. While the energy per quantum is normally sufficient in the

(a)

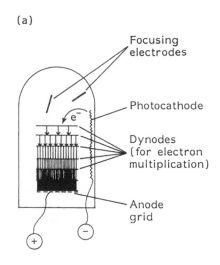

Focusing electrodes

Photocathode

Dynodes (for electron multiplication)

Anode grid

(b)

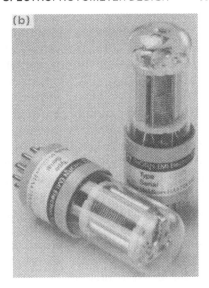

FIGURE 3.7: A photomultiplier tube. **(a)** Diagram showing amplification as photoelectron is accelerated, knocking several electrons from a dynode followed by a repeat of this process at subsequent dynodes. **(b)** A suitable photomultiplier. Side window photomultipliers are convenient for spectrophotometry as the slit image can be focused directly on the window. Photograph courtesy of Thorn-EMI Electron Tubes Ltd.

ultraviolet and at the blue end of the visible spectrum, it can be too small at the red end; thus the response of photomultipliers tends to fall off towards the red/near infrared.

Both of these drawbacks can be overcome by using photodiode detectors. These are solid state devices (based on silicon) whose resistance to current flow depends on the intensity of light falling on them. (Some commercial spectrophotometers use a photomultiplier below 600–650 nm, but switch to a photodiode above this wavelength.) Although photodiodes may be somewhat less sensitive than photomultipliers, their small size and low voltage requirements makes them attractive for portable instruments and allows them to be built into arrays (diode array detectors) for sampling multiple beams (multiple wavelengths) simultaneously (see Section 4.4).

3.6 Varying the wavelength (scanning)

For measuring absorption or fluorescence spectra (the variation of absorbance or fluorescence with wavelength), the wavelength of the light falling on to the sample must be varied. This is typically achieved by rotating the grating within the monochromator, so that a different region of the dispersed beam falls on the exit slit. The system is calibrated so that the position of the grating in its rotation directly signals the wavelength of emerging light.

In the simplest case, the grating is driven by a geared wheel which is turned by the operator; the wheel is marked off in wavelength units, so that its position relative to a reference mark gives the selected wavelength. This allows individual measurements to be taken at a number of separated wavelengths, and thus a spectrum to be constructed. However, it does not allow a smooth variation of wavelength and measurement of a continuous spectrum. This is achieved by driving the grating using an electric stepping motor (which allows a small, fixed period of observation in each position), while both the position of the grating and the light reaching the detector is continuously monitored. Modern instruments can scan across a spectrum in this way at up to 40 nm sec^{-1}.

As has been noted previously, even the best lamps do not emit light equally at all wavelengths. Thus, during a wavelength scan, the energy reaching the detector will vary with wavelength not only due to differential absorption by the sample (which is useful information) but also due to differential emission by the lamp (which is not). Modern spectrophotometers commonly compensate for differential emission either mechanically, by automatically varying slit width with wavelength, or electronically, by varying the gain on the photo-multiplier with wavelength. These devices ensure that the perceived signal does not vary with wavelength in the absence of sample, and thus that any changes observed do indeed reflect the properties of the sample under study. Nonetheless, all such correction systems have imperfections, and it is always good practice to measure a 'spectrum' in the absence of sample for comparison with that of the sample itself.

Scan rates faster than 40 nm sec^{-1} are difficult to achieve by mechanically driving the monochromator. An alternative approach is to measure simultaneously over a range of wavelengths, using a series of diodes in a linear arrangement ('diode array'). In this case, rather than illuminating the sample sequentially with light of different wavelengths, the sample is illuminated directly with white light from the lamp, and the wavelengths dispersed (by a grating) only after the beam emerges from the sample (see Chapter 4, *Figure 4.7*). Each wavelength thus impinges on a different region of the diode array, and an entire spectrum can be measured instantaneously. This optical arrangement, where the beam is dispersed after it has passed through the sample, is known as 'reverse optics' in contrast to 'conventional optics' where the sample lies after the monochromator.

3.7 Data capture

Except for instruments working at very low light levels, the signal from a photodetector in a photometer or fluorometer will be an electric current – an analog signal. Historically, this has been measured by using this current to deflect an inbuilt galvanometer (for constant signals), to displace a pen on an *x-t* recorder (where the half time for the change in signal lies between 5 sec and 1 h), or to displace an electron beam in an oscilloscope (for signals changing more rapidly).

Nowadays, a static signal is commonly read by an inbuilt digital (LED or liquid crystal) display, calibrated in absorbance (or transmittance) units. While these are convenient, care must be taken to avoid 'spurious accuracy' – just because a digital display can be read to three decimal places, a reading of '0.357' does not imply confidence with the last figure and usually means no more than 'between 0.35 and 0.36' as was read from the older galvanometer instruments.

Output which varies in time (as a result of wavelength scanning or chemical changes in the sample) is today increasingly passed through an analog to digital (A/D) converter and stored on a computer, either a dedicated computer built into the spectrophotometer (the option preferred by most manufacturers) or a stand-alone personal computer. With suitable software this allows, for example, facile calculation of rates, statistical analyses of data, and addition, subtraction and correction of spectra. While this facility has great potential, readers should note that the software available with any given instrument may not be especially user-friendly, nor may it allow easy transfer of data to other computers. The volatility of electronic data also needs consideration, and regular back-up copies should be made.

3.7 Data Capture

4 Geometry, Light Paths and Beam Splitting

4.1 Sources of error in absorbance measurements

4.1.1 Variation in wavelength response

One problem in spectrophotometer and fluorometer design is due to the variation in lamp output, and the response of the detector, with wavelength. In Chapter 3, we saw how this problem could be reduced by a suitable choice of lamp and detector type. Electronic compensation may also be employed.

Alternative approaches to this problem involve refining the optical design of the instrument to incorporate two light beams, a measuring beam and a reference beam. Any variations in response to the measuring beam are matched by parallel variations in response to the reference beam, and so they can be eliminated from the data by subtraction. This arrangement occurs in split beam or dual wavelength instruments, which are discussed below.

Other problems encountered in precise and consistent measurement of absorbance and fluorescence have also influenced spectrophotometer design.

4.1.2 Noise

Noise arises from random fluctuations in lamp intensity, and occasional (background) discharges in the photomultiplier. Both are predominantly electrical in origin.

Noise is observed as random variation around a mean signal. If the noise level is high and/or the signal small (a low signal/noise ratio), then the noise may obscure the signal or make its measurement inaccurate. Note that a small signal is observed when only a small amount of light reaches the detector – a low fluorescence, or a high reading in absorbance measurements. Such effects may be reduced by electronically stabilizing the lamp and detector outputs, as in the Beckman 600 series. Alternatively, split beam or dual wavelength optics will decrease the effect of lamp fluctuations on the photometric measurement.

4.1.3 Stray light

Stray light is light arriving at the detector without its having interacted correctly with the sample. This may be because it is of an inappropriate wavelength. This would typically represent light selected from higher order reflections by the grating. Alternatively it may have bypassed the sample through reflections in the sample compartment or cuvette, due to incorrect alignment of the sample in the light path, or it may arise from entry of light from outside the instrument (*Figure 4.1a*).

Stray light spuriously increases the signal from the detector. Again it becomes more significant if the true signal is small, for example at high optical densities. This can be seen by considering the expression for absorbance

$$A = \log (I_0 / I)$$

If stray light i falls on the detector we have

$$A_{app} = \log (I_0 + i)/(I + i) \qquad (4.1)$$

which reduces to $\qquad A_{app} = \log (I_0 + i)/ i \qquad$ where I is small

and hence $\qquad A_{app} < A$

The effect of stray light on absorption measurements is shown in *Figure 4.1b*. Stray light is reduced (i) by using filters to eliminate higher order reflections from the grating, and (ii) by using lenses to focus the beam on to a small area of the sample. If the beam is focused very precisely, very small samples (down to 5 µl) can be measured. Typical values quoted for stray light in instrumental specifications are less than 0.1% transmittance; at this level stray light has little effect at low absorbance values, but at $A=3$, from Equation 4.1, $A_{app}=2.69$, giving an error in the measured absorbance of about 10%.

(a)

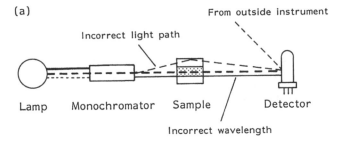

(b)

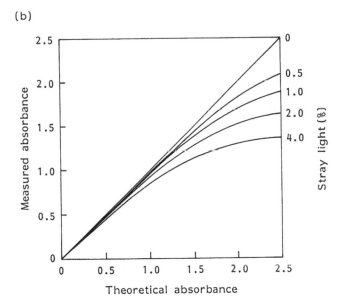

FIGURE 4.1: *(a) Sources of 'stray light'. (b) The effect of 'stray light' on measurement accuracy.*

4.1.4 Light scattering (turbidity)

Many biological samples comprise a suspension of small, insoluble particles such as bacterial cells, or membrane fragments, and are thus not optically clear (see Chapter 1). In such samples the chromophores absorb light, but in addition the particles scatter it. In absorbance measurements, therefore, less light arrives at the detector than would be expected on the basis of absorbance alone.

(a)

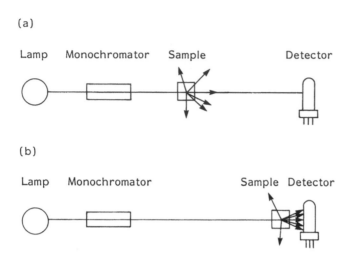

FIGURE 4.2: *Effect of sample cell position with turbid samples. (a) Most light is scattered away from the detector and thus the signal intensity is low. (b) The sample cell is placed close to the detector and a significant fraction of the deflected light still reaches the detector.*

This leads to two problems in measurement. First, the overall loss of light reaching the photomultiplier will lead to a lower signal/noise ratio (see Section 4.1.2). This can be obviated to some extent by placing the sample close to the photomultiplier, so that much of the deflected (scattered) light still reaches the detector (*Figure 4.2*). Failing this, if the sample is highly turbid or opaque, the amount of reflected light can be measured in a reflectance spectrophotometer.

More insidious are the effects of changes in turbidity over time. Suspensions of biological material are notorious for changing their scattering properties during the course of an experiment. This is often due to settling of the material, but also may result from denaturation, or osmotic changes in volume. These effects can be compensated for by using two measuring beams, one at a wavelength absorbed by the chromophore of interest and a second to monitor light scattering, at a wavelength removed from that absorbed by the chromophore. Instruments with the requisite optical configuration are known as dual wavelength spectrophotometers.

4.1.5 Spectral bandwidth

In Chapter 1 we noted that the absorption spectra of compounds in solution show relatively broad absorption bands. Strictly, we can define the natural bandwidth as the width across an absorption peak at half its maximal height (*Figure 4.3a*) – typically 60 nm (for NADH, as an example) in biochemical systems. The natural bandwidth is a property of the sample, and is independent of the method of measurement.

To measure absorption bands, the wavelength of illumination is varied and the varying absorption measured. However, we cannot illuminate a sample with light of one single wavelength; the range of wavelengths used for illumination is defined by the spectral bandwidth of the illuminating beam, which is half the width of the band defined by the exit slit. Thus the absorption measured is effectively an 'average' across this narrow band of wavelengths.

Clearly, if the spectral bandwidth (for the illuminating beam) exceeds the natural bandwidth (of the absorption band), the measured absorption spectrum will differ from the true spectrum. The measured peak will be too broad and its height reduced, due to this averaging process. The observed spectrum will be distorted in relation to the true spectrum (*Figure 4.3b*).

To avoid this distortion (for good resolution), the spectral bandwidth must be small compared with the natural bandwidth. Spectral bandwidths of less than 15% of the natural bandwidth are acceptable, for example, giving errors of less than 1% in the measured peak height. For accurate measurement, therefore, spectral bandwidth should be decreased to that level by decreasing the slit width (see Chapter 3) at the exit of the monochromator.

However, too great a decrease in slit width will cut down the amount of light reaching the detector, leading eventually to a poor signal/noise ratio as outlined above. Thus the user must adopt a compromise between bandwidth and signal intensity acceptable for his particular purpose. Clearly, the more powerful the lamp used, the narrower can be the spectral bandwidth without too low an energy reaching the detector – high lamp intensity promotes high spectral resolution.

In fact, slitwidths are not under the control of the user in most modern commercial instruments, as noted in Chapter 3 (although they may be varied automatically with wavelength). In this case, the manufacturer's specification (typically corresponding to a spectral bandwidth of <2 nm) should be noted.

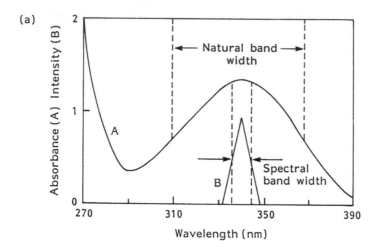

(a)

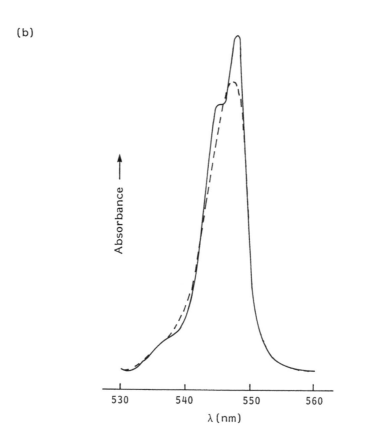

(b)

4.2 The single beam spectrophotometer

This is the simplest type of spectrophotometer. A single light beam passes through a single measuring position, falling on a single detector. To compare a sample with a reference cuvette, the reference is first placed in the measuring position, measured and then replaced by the sample – the two must be measured sequentially.

The optical diagram of such an instrument is very similar to that of the idealized instrument shown in *Figure 3.1a*, containing lamp, monochromator, sample compartment and detector. Wavelength selection normally occurs between lamp and sample (*Figure 4.4*), although in some cases selection may be between sample and detector, a situation known as reverse optics. As compared with *Figure 3.1*, real instruments also contain a number of mirrors and lenses (i) to focus light precisely on to the sample, (ii) to allow the user to switch between deuterium (ultraviolet range) and tungsten–halogen (visible range) lamps, and (iii) to keep the instrument compact (small bench footprint).

Variations on this theme may include the addition of a mechanical 'chopper', a rotating disc which is pierced to interrupt the beam at intervals (Uvikon Series) or a pulsed light source (Pharmacia Ultrospec), which allow the use of a.c. amplifiers in the detection circuit (see Section 3.2). A filter may also be included at the exit of the monochromator to cut out higher order reflections.

Single beam instruments are useful for routine biochemical arrays, where repeated, static measurements at a single wavelength are required. It is clearly important in such instruments that the light

FIGURE 4.3: *(a) The relationship between natural and spectral bandwidths. The natural band width shown is for NADH (58 nm, curve A). Curve B is a schematic representation of the spectral bandwidth of the monochromator exit beam. Reproduced with permission from Gerhardt (ed.) Figure 5 in* Manual of Methods for General Bacteriology, *9th edn. © 1991 with permission from the American Society for Microbiology, Washington. (b) Effect of spectral bandwidth on (reduced–oxidized) spectrum of cytochrome c. (——) spectral bandwidth 0.5 nm; (---) spectral bandwidth 3 nm. The spectra were taken at 77 K. Note the loss of resolution and height with a broad spectral bandwidth. Data from Poole and Bashford (1987)* Spectrophotometry and Spectrofluorimetry: a Practical Approach *(D.A. Harris and C.L. Bashford, eds) pp. 23–48.* IRL Press, Oxford.

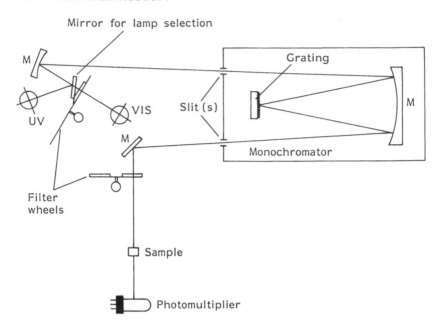

FIGURES 4.4: *Single beam spectrophotometer using conventional optics (Unicam PU 8700 Series).*

and detector outputs be stable, as fluctuations in the light source would lead to fluctuations in the perceived measurement. In the Beckman 600 series, the lamps and detector are stabilized electronically (so-called 'stable beam technology'), whereas other manufacturers rely on variations in the optical arrangement to compensate for such fluctuations (see below).

4.3 The double beam spectrophotometer

In the double beam spectrophotometer, the illuminating beam is split so as to fall on both reference and sample cells. The signal then detected represents the difference between sample and reference. This arrangement automatically corrects for lamp fluctuations, since they affect sample and reference signals equally. In addition it obviates the need for the operator, or the instrument, to store the reference signal (i.e., the spectrum of buffer alone) and subsequently subtract it from the signal obtained with the sample under study.

Double beam spectrophotometers use a single monochromator to select the wavelength of the beam. Classically, they have also used a single detector, with the beam being switched from sample to

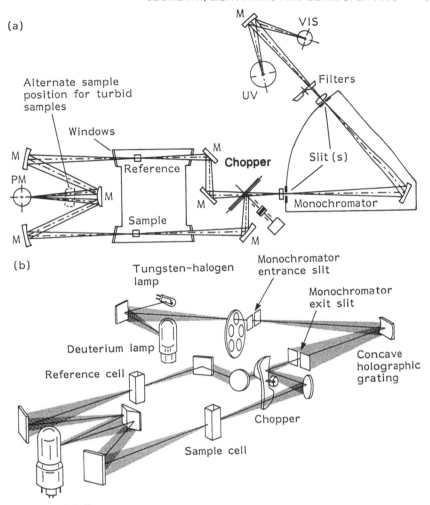

FIGURE 4.5: *(a) Double beam spectrophotometer (Uvikon 900 Series, Kontron Instruments); (b) a three-dimensional representation of (a).*

reference cuvette by a rotating mirror – again known as a chopper (*Figure 4.5a*). This switching occurs at the rotation frequency of the chopper (50–500 Hz). After passing through either cuvette, light is focused on the same area of the detector, to ensure that each beam is monitored with the same sensitivity. A signal from the chopper instructs the electronics which beam is falling on the detector at any instant, and the reference and sample beams can thus be compared. This is shown in the three-dimensional representation in *Figure 4.5b*.

An alternative approach splits the illuminating beam with a half silvered mirror, so that half the energy passes simultaneously through both reference and sample. In this case, separate detectors must be used for both reference and sample chambers. This arrangement has a potentially higher kinetic resolution than the classical instrument of *Figure 4.5*, as its time resolution is not limited by the speed of the chopper. However, its optical resolution may be compromised by difficulties in providing two detectors with identical response characteristics ('well matched' detectors). Both of these types of spectrophotometer may be referred to as 'double-beam', 'dual beam' or 'split beam' instruments – while the latter term is more descriptive, the first is in wider use. Unfortunately, there is no agreed convention for naming instruments that use separate beams for sample and reference cells.

4.4 Variations on a theme

4.4.1 The ratio spectrophotometer

The double beam spectrophotometer compensates for variations in lamp intensity by measuring a reference beam split off from the sample beam, after it has passed through a reference cuvette. A similar compensation can be made by simply splitting off some light from the sample beam and leading it directly to a detector. Such an instrument (*Figure 4.6*) has a single sample compartment (no reference cell) like a single beam instrument, but measures the ratio of the signals at the two detectors – it is a ratio spectrophotometer. Such instruments give high performance for fixed wavelength measurements but will not, of course, automatically subtract the spectrum of a reference from a sample cell as do double beam instruments.

4.4.2 Microplate spectrophotometer

Ratio spectrophotometry is also employed in a new breed of spectrophotometers designed for rapid measurement of multiple samples – microplate spectrophotometers. In all the instruments described so far, the measuring beam(s) passes horizontally through the sample contained in a flat, parallel sided cuvette. This limits the number of samples which can be easily handled to four, six or at the most 12 with conventional sample holders.

The microplate reader (e.g. Molecular Devices SPECTRA$_{max}$™) typically uses a plastic plate containing 96 wells as is commonly used for immunoassays. The light passes vertically through the sample, that is, through the air/liquid interface and out through the base of the plate. Unlike conventional cuvettes, the optical path length

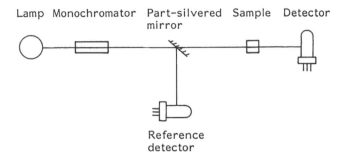

FIGURE 4.6:   *Basic structure of a ratio spectrophotometer. Part (~10%) of the light is diverted, so that the lamp fluctuations can be recorded by the reference detector.*

depends on the volume of liquid in each well, but provided this remains constant, a rapid comparison of the absorbance of a large number of samples can be measured.

Microplate instruments typically have two photodetectors, one acting as the reference. The microplate is then mechanically driven above the other detector and each well measured in turn.

4.4.3 Diode arrays for instantaneous spectrum measurement

In Chapter 3, we noted the possibility of using a diode array as detectors to measure a number of wavelengths simultaneously. An optical diagram of such an instrument is shown in *Figure 4.7*. Such an instrument will always use 'reverse optics' (dispersing the beam after passing through the sample) and, as yet, only the single beam arrangement is commercially available.

4.4.4 Fiber optic spectrophotometers for inaccessible samples

In the above instruments, the various optical components must be precisely aligned relative to the sample holder on a fixed optical bench. This arrangement is bulky, and precludes the study of samples located in restricted environments (e.g. inside a large magnet, in magnetic resonance spectroscopy). Such samples can be studied, however, by coupling the light source to the detector through flexible optical fibers.

Fiber optic spectrophotometers are typically modular and use reverse optics for reasons of space. They are centered around a light tight

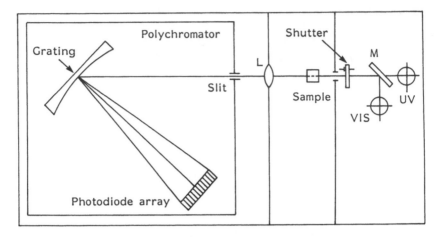

FIGURES 4.7: *Spectrophotometer with diode array detector (Beckman DU 7000 Series). Note reverse optics arrangement.*

sample holder, with associated collimating lenses, made in-house to fit the environment of study. This holder bears standard (SMA) fiber optic couplings. White light is led to the sample from a stabilized light source (e.g. a quartz–halogen lamp) via a single core quartz (UV quality) optical fiber of 200–500 μm diameter. The light from the sample, emerging through a similar fiber, is analyzed by a grating (400 lines per mm) and a photodiode array (Oriel). Light falling on the array generates the electrical signal which, when digitized and fed to a suitable computer, will generate a spectrum, or if required, follow a single wavelength with time. Using two detectors, a ratio or double beam mode of operation can be operated.

4.5 The dual wavelength spectrophotometer

In the dual wavelength spectrophotometer, a single sample is illuminated with light of two different wavelengths, and the signal taken as the difference between the two absorbances. Typically, one wavelength is taken as the 'reference' wavelength, for example, the isosbestic wavelength of the compounds under study, while the other wavelength is the 'measuring' wavelength, at which changes in absorbance occur in the chromophore.

Light at two different wavelengths must therefore be generated simultaneously, and dual wavelength instruments, in contrast to

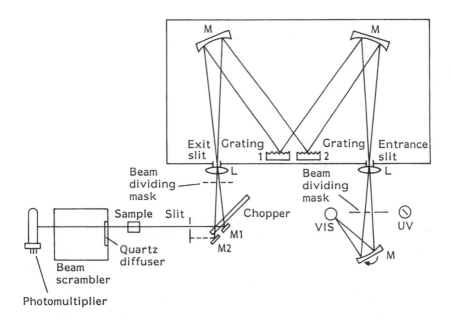

FIGURE 4.8:   *Dual wavelength spectrophotometer (SLM-Aminco DW Series). Light from grating 1(λ_1) is directed to mirror M1. Light from grating 2(λ_2) is directed to M2. The light from each mirror accesses the sample alternately as the chopper rotates.*

double beam instruments, have two monochromators (*Figure 4.8*). Light from each monochromator is focused on the same region of the sample by a rotating chopper (see *Figure 4.5*), so that the illumination switches between the two wavelengths at 50–500 Hz. As before, a signal from the chopper informs the single detector which wavelength is operative at any instant. [An alternative arrangement is to employ two detectors, with the two illuminating beams passing orthogonally through the sample cell; while this has better kinetic resolution, such instruments are rare in biochemistry laboratories.]

Since, in dual-wavelength instruments, the reference and measuring beams pass through the same sample, such instruments are useful if the sample undergoes nonspecific changes which may not be reproduced in a separate reference cuvette – in particular if light scattering changes in turbid samples. The reference beam, then, effectively monitors any turbidity changes while the measuring beam monitors true absorbance changes.

As noted above, turbidity (in suspensions of membrane vesicles or cells) also causes problems by scattering light away from the photodetector, and thus decreasing the intensity of the measuring beam. As a result, dual-wavelength spectrophotometers are configured so that the sample holder is close to the detector (compare *Figure 4.8* with *Figure 4.2*), an adaptation to reduce signal loss from turbid samples (see Section 4.1.4).

4.6 Sources of error in fluorescence measurements

The basic components of a spectrofluorometer are the same as those of a spectrophotometer, namely lamp, monochromator, sample compartment and photodetector. Design considerations are also similar, including optimizing the signal/noise ratio and minimizing the effects of stray light, turbidity and (when measuring spectra), spectral bandwidth. However, the specific properties of fluorescence emission have led to rather different solutions in spectrofluorometer design from those in spectrophotometers.

4.6.1 Signal/noise ratio

In absorbance measurements, a high proportion of light from the illuminating beam falls directly on the detector – in routine measurements certainly more than 10% (an absorbance below one unit corresponding to more than 10% of incident light transmitted). In fluorescence measurements, only light emitted by the sample itself is measured, with an intensity typically less than $1/10^6$ of the incident beam. This has led to the routine use in fluorometers of more intense lamps, such as xenon arc lamps, despite their less favorable spectral emission characteristics. Pulsed light sources are common. Filters, which typically pass more light than monochromators, are also used for wavelength selection in standard laboratory fluorometers while they rarely appear in modern spectrophotometers. In addition, photon counting detectors are more common in fluorometry (e.g. PTI-RF Series).

Paradoxically, it is the elimination of the light from the illuminating beam which gives fluorescence measurements their inherently higher sensitivity than absorbance measurements (i.e. their ability to detect very small amounts of material). A small amount of material in an absorbance measurement will lead to only a small percentage fall in the light transmitted; in effect the detector needs to detect a small difference between two intense light beams. A small amount of material in a fluorescence measurement, however, will increase the light falling on the detector by a very large percentage – since in the absence of the compound, no light will reach it (see Section 2.5). Large percentage differences are more easily measured than small ones, and the resulting signal is greater in a fluorescence measurement. If the noise level is low (intense lamps etc.), a good signal/noise ratio will be obtained.

4.6.2 Stray light and turbidity

Since the light emitted is so much lower in intensity than the illumination beam, it is essential in fluorometers to ensure that the illumination beam, or its reflected or scattered components, do not reach the detector. Such exclusion can be maximized by placing the detectors at right angles to the incident beam, and this orthogonal arrangement is common in commercial instruments. However, even with optically clear solutions, Rayleigh and Raman scattering (see Chapter 1) contribute to the fluorescence signal obtained, and need to be allowed for. With turbid solutions, the contribution of scattering effects rapidly becomes overwhelming, and the geometry of the system may have to be changed in order to measure only fluorescence from the front of the sample ('surface' or 'front face' fluorescence).

4.6.3 Spectral bandwidth

Because of the relatively low intensity of emitted light, the compromise between signal/noise ratio and spectral bandwidth is even more marked in fluorescence than in absorbance measurements.

This, it turns out, is rarely a problem in fluorescence measurements on biological systems. Firstly, fluorescent emission bands tend to be broad so that even a relatively wide spectral bandwidth causes little distortion in their measurement. Secondly, far fewer compounds are fluorescent that simply absorb light (see Chapter 2), and thus, whereas overlapping absorption bands are quite common, even measurements made using a wide spectral bandwidth are unlikely to show interference from other fluorophores in the same sample.

4.7 The spectrofluorometer

Laboratory spectrofluorometers are almost invariably analogous to the single beam spectrophotometer, with a single illumination beam passing through a single sample. Some instruments (e.g. Perkin-

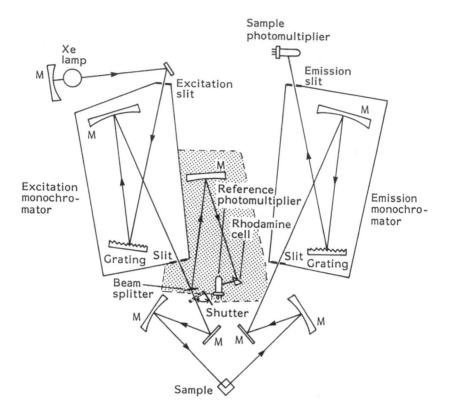

FIGURES 4.9: *Single beam spectrofluorometer (Perkin-Elmer LS Series). Note the roughly symmetrical organization with separate excitation and emission monochromators. The shaded region shows the beam splitter/reference photomultiplier included to correct for lamp fluctuations in the LS5 models.*

Elmer LS5) do allow a ratio mode of operation, which corrects for fluctuations in the light beam; here a small fraction ($\approx$10%) of the incident beam is diverted to a second (reference) detector (*Figure 4.9*). Besides right angle geometry, the most obvious difference between a spectrophotometer and fluorometer is the presence, in the latter, of two monochromators. One selects the wavelength of illumination, while the second selects the wavelength at which emission is measured. As noted previously, light is emitted from a fluorophore at a higher wavelength than the exciting light. These two monochromators can be scanned independently to measure either an excitation spectrum (which should be identical to the absorption spectrum), or an emission spectrum (Chapter 1, Section 1.7).

Another distinguishing feature is that the exit slits from the two monochromators, in a fluorometer, are normally accessible to the user and hence their width can be adjusted manually. (These are termed the excitation slit and emission slit respectively.) This allows fluorescence to be measured over a much wider range of concentrations (up to 10^6-fold difference) than can absorbance. For fixed wavelength measurements, it is convenient to keep the excitation slit narrow, and use a broad emission slit for increased intensity (since emission bands are generally broad). For measurement of spectra, considerations of spectral bandwidth, the class of spectrum to be measured (emission or excitation), and signal intensity will influence the choice of slitwidth.

4.8 The dual wavelength fluorometer

The fluorescence of a probe molecule varies with its environment, resulting in both a change in fluorescence intensity and in the position of the absorption/emission peaks (see Chapter 2). Measuring amounts of material or the behavior of probes in dilute solution, it is convenient to follow fluorescence intensity, using a single beam fluorometer.

When dealing with probes within cells, however, it is more convenient to follow spectral shifts. The observed changes are then independent of the amount of probe in the light beam, and thus unaffected by, for example, changes in cell thickness or destruction of the probe by metabolism. Best results are obtained when measurements are made at two wavelengths, each characteristic of one form of the probe (e.g. protonated and unprotonated). This then requires a fluorometer capable of excitation at two wavelengths, or of measuring emission at two wavelengths. It is conventional to calibrate the measurements in terms of the ratio of (emission at λ_1)/(emission at λ_2) and thus such instruments are known 'confusingly' as ratio fluorometers.

Since each wavelength requires its own monochromator (or band pass filter), a dual wavelength fluorometer can possess up to four monochromators – two each for excitation and for emission. This is costly, and generally the user decides on one configuration suitable for his application. For example, measurement of intracellular Ca^{2+} using indo-1 uses two emission wavelengths, with excitation at a single wavelength; with fura-2, one emission wavelength is followed after excitation at two different wavelengths (*Figure 4.10*).

The optical arrangement is simple when two emission wavelengths are studied; each is placed orthogonal to, and on opposite sides of, the

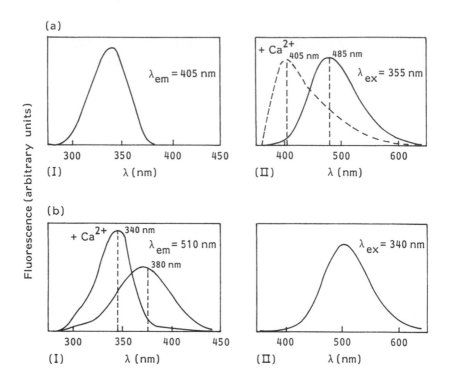

FIGURE 4.10: (a) Excitation (I) and emission (II) spectrum of indo-1 in the presence or absence of saturating [Ca²⁺]. The excitation spectrum is hardly affected by [Ca²⁺]. Measurements of [Ca²⁺] are made by comparing emission at 485 nm and 405 nm. (b) Excitation (I) and emission (II) spectrum of fura-2 in the presence or absence of saturating [Ca²⁺]. The emission spectrum is hardly affected by [Ca²⁺]. Measurements of [Ca²⁺] are made by comparing emission after excitation at 380 nm and 340 nm.

cuvette. For measurement with two excitation wavelengths, a chopper is used to separate the two beams in time. In the PTI ratio fluorometer, the chopper is placed before the monochromators (compare *Figure 4.11* with *Figure 4.6*) so that the energy is not divided between them; each monochromator in turn receives the full energy of the lamp, maximizing the fluorescence output. In this instrument, the two wavelengths are combined by a bifurcated single fiber optic before the sample. With this method the time resolution of measurement is limited by the rotation rate of the chopper (up to 500 Hz).

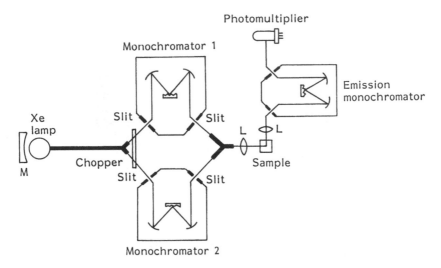

FIGURE 4.11: *'Ratio fluorometer' using two excitation wavelengths (e.g. for fura-2 measurements) (Photon Technology International). Photon counting detection is often used for high sensitivity (see text).*

4.9 Measurements on opaque samples

It is difficult to measure either absorbance or fluorescence in samples which scatter significant amounts of light. In absorbance measurements, scattered light is prevented from reaching the detector, and the apparent absorbance is too high. In (right angle) fluorescence measurements, light is scattered towards the detector, and the apparent fluorescence is, similarly, too high.

In (concentrated) highly colored solutions of fluorophores, a second effect – the inner filter effect – may also distort the fluorescence reading. This occurs when so much of the excitation light is absorbed that the distal region of the cuvette is poorly illuminated. Fluorescence from this region will thus be less intense than from the proximal region of the cuvette. This can be a particular problem if the detector samples only the central region of a square cuvette (*Figure 4.12*). In a homogenous solution, this problem can be overcome by dilution; however, this may not be possible, for example, in tissue samples or with macromolecular assemblies such as actomyosin filaments.

(a)   (b)

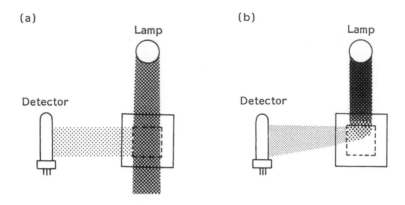

FIGURE 4.12: *The inner filter effect. (a) In a dilute solution only a very small fraction of incident light is absorbed, and all parts of the sample receive excitation light. (b) In a concentrated solution most of the incident light is absorbed before it traverses the cuvette. Distal regions of the sample do not receive excitation light and thus fluorescence is sub-optimal.*

These problems can be alleviated by measuring light emerging from the front side of the sample – either reflected light (in reflectance spectrophotometry) or emitted light (in front face fluorometry). The detector must now be located on the same side of the sample as the lamp (see below). While such instruments are relatively straight-forward to construct, it is notable that the distance the light penetrates into the sample in these instruments is not known, and so an absolute calibration is not possible. These instruments must always be calibrated relative to known standards of the material under study.

An example of a commercial reflectance spectrometer is the Boehringer Reflotron® (*Figure 4.13*). Light is emitted from a light-emitting diode and is detected at photodiode detectors situated within a white walled (Ulbricht) sphere. A light emitting diode emits only a narrow band of wavelengths, and so no monochromator is required. To follow a range of chemical reactions, therefore, a range of light-emitting diodes is required; the three available with this instrument emit around 567 nm (green), 642 nm (orange), and 951 nm (far red).

Within the sphere, one detector is situated close to the sample to be measured, and receives light reflected from the sample (I_s); the other (reference) detector is symmetrically placed within the sphere, but receives light simply reflected off the walls (I_0). If the sample absorbs

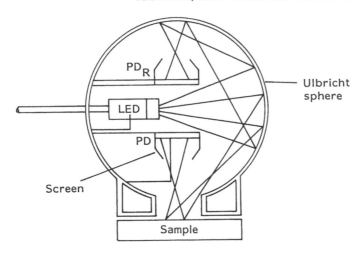

FIGURE 4.13: *Optical arrangement of a reflectance spectrometer (Reflotron®, Boehringer Mannheim). LED, light-emitting diode (fixed wavelength); PD, photodiode detector, PD$_R$ reference detector. The sample is introduced on a paper/plastic strip.*

light at the emitted wavelength, less will be reflected and a decrease in the I_s/I_o ratio will be recorded.

This instrument is typically used in clinical practice. The test material (typically a drop of blood) is applied to a paper stick on which the necessary reagents have been absorbed; the chemical reaction produces a colored compound and this colored region of the stick is then measured by the photometer. Note that only the sample is applied in solution – all the other components necessary for reaction are already present in the stick (solid phase), which can be thrown away after use. This ease of operation allows the use of this technology by unskilled operators in relative safety (no solutions to make, no liquids to dispose of) and such instruments are thus convenient for clinical measurements, for example for measuring blood glucose in diabetics, or blood cholesterol.

For front face fluorometry, many commercial spectrofluorometers can be fitted with a front face accessory which (with the aid of suitable mirrors) will modify the sample holder suitably. Note that, for such measurements, the angle between the excitation and emission beams is typically less than 90°, to minimize interference from reflected light. For solid samples, such as biological tissues, light can be led to and away from the tissue using a bifurcated fiber optic fitted to suitable filters or monochromators.

5 Measuring Absorbance and Fluorescence – Aspects of Instrument and Sample Preparation

5.1 Calibrating the instrument

Before making any measurements using a spectrophotometer, we must know if it is working correctly. Three points should be checked:

(i)   the dark current: the electronics should be adjusted so that no current flows in the detector when no light reaches it ($A = \infty$);

(ii) wavelength calibration: the monochromator must be correctly aligned so that the wavelength displayed is actually the wavelength illuminating the sample;

(iii) absorbance calibration: the response of the detector, in particular its slope (change in response/change in signal), must be such that the absorbance is correctly assessed.

In most modern instruments, these three checks and the relevant adjustments are made automatically when the instrument is switched on, using a resident program. The operator needs only to note any error messages. Of particular interest is any wavelength adjustment deemed to be necessary: if this is recorded as more than a few nanometers, the optics may well be misaligned due to, for example, mechanical shock, and the instrument will need workshop attention. Where these checks need to be made manually – or if the automatic calibration is suspect – the following procedure should be followed:

(i)   With the instrument on, the light path between lamp and detector should be interrupted using a shutter, by closing the slit completely or by interposing an opaque (black metal) plate. The instrument should then be adjusted so that the reading obtained is set to zero transmittance/ infinite absorbance.

(ii)   The internal wavelength calibration normally utilizes the bright line of the deuterium lamp in the visible region (λ_{max} = 656.1 nm) as a standard. For an external calibration/check it is usual to interpose, in place of the sample, a glass doped with oxides of the rare earths, holmium or didymium. The absorption spectrum of this glass is then measured, and the observed position of the peaks checked against the known values. Holmium, for example has nine sharp maxima between 240 and 650 nm (*Figure 5.1*), and didymium five between 570 and 810 nm (not shown). Conflict between the known values for these maxima and the measured values requires the wavelength calibration to be adjusted as described in the manufacturer's instructions.

(iii)   To calibrate absorbance it is possible to insert neutral density filters (Kodak–Wratten) into the sample position. These are available with an optical density of 1; one such filter in the light path should give $A = 1$, two should give $A = 2$. Use of these filters also allows the operator to check for stray light; the measured absorbance drops below the expected value as A increases if stray light becomes significant (see Section 4.1).

An alternative calibration can be made using a 0.2 mM solution of potassium chromate in 0.05 M potassium hydroxide, which can readily be made up quantitatively. This solution should have A_{275} = 0.736 and A_{375} = 0.964.

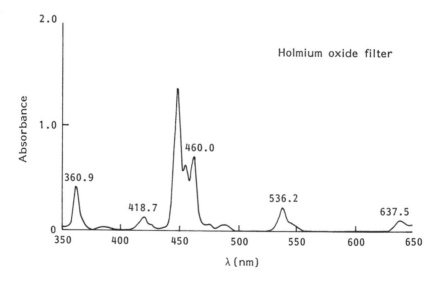

FIGURE 5.1:   *Spectrum of holmium oxide filter in the visible region, showing position of peaks useful for calibration.*

On an instrument which is not self calibrating, the dark current must be set whenever the instrument is switched on. Wavelength and absorbance calibrations, however, are required only periodically providing the instrument is maintained in a fairly constant environment. It is useful to record calibration data so that performance of an instrument can be monitored over time.

5.2 Choosing the cuvette

5.2.1 Size

Sample compartments supplied with commercial spectrophotometers are normally designed for cuvettes of 1 cm × 1 cm square base. The simplest design, therefore, is the square cuvette with a central cavity 1 cm × 1 cm x 4.5 cm (see *Figure 3.6*), which has a working volume of 2–3 ml.

If only smaller volumes of solution are available, a common solution is to retain a 1 cm light path but decrease the width of the internal cavity to, say, 4 mm, giving a working volume of about 0.7 ml. Such 'semimicro' cuvettes are available with blackened ('masked') walls, to prevent light reaching the detector without passing through the sample itself, and to avoid reflection off the internal walls.

If the light beam is focused precisely, even smaller capacity cuvettes can be used, culminating in the ultramicro 5 µl cuvette available for the Beckman D series. In these cases, an adapter must be fitted to the normal sample holder (or this sample holder replaced by another) to align the cuvette precisely in the light beam.

While the volume of sample available is commonly the major factor in determining choice of cuvette size, it may be desirable to change the pathlength for either very concentrated (high absorbance), or very dilute (low absorbance) solutions. Cuvettes with pathlengths from 1 mm to 100 mm are available (Hellma), again provided that a suitable sample holder is available.

5.2.2 Material

The choice of cuvette material is governed by (i) the wavelength range under study, (ii) the temperature range used, and (iii) cost. For room temperature measurements, plastic (polyacrylate) cuvettes have minimal absorption down to 250 nm (see *Figure 3.5*), and are adequate for most visible and near ultraviolet measurements in biochemistry. Only silica (quartz) is suitable for measurements down to 170 nm (for UV), although care must be taken that a suitable grade of silica (UV-grade silica, Spectrosil) is obtained, as some grades absorb significantly above 220 nm.

For applications where temperature control is critical, glass or quartz cuvettes are preferred since the thermal conductivity of plastic is very poor. Optical glasses, however, have significant absorption below 320–350 nm (depending on manufacturer), so for such measurements, silica cuvettes are required even for measurements in the near UV region. It is, of course, possible to use silica cuvettes in the visible region in place of glass, but since the cost of silica cuvettes is some threefold higher, and breakages equally frequent, glass cuvettes are generally used!

For work with frozen solutions, perspex (polymethacrylate) cuvettes are convenient, as they can withstand repeated cooling/warming cycles and, when they do eventually crack, they are relatively cheap to replace.

5.2.3 Position of light beam

For meaningful absorbance readings, light should pass linearly through the sample. Furthermore, no light should reach the detector which has not passed through the sample. This requires that the sample cuvette (i) must be correctly aligned in the light beam and (ii) must be filled to an appropriate level.

Spectrophotometers are typically aligned so that a glass cuvette 1 cm × 1 cm × 4.5 cm (internal dimensions) is adequately illuminated when filled to within 1 cm of the top (*Figures 5.2*). The problems that may arise with narrower widths of cell have been noted. They arise due to internal reflections and/or light passing through the outside walls of the cell, and can be alleviated by using masked (black walled) cuvettes (see Chapter 3, *Figure 3.6*) or by interposing a (black metal) mask either close to the monochromator (effectively narrowing the slit width and hence the width of the illuminating beam) or close to the cuvette. The extent of this problem depends on how far the instrument collimates the illuminating beam or focuses it on to the sample (see Chapter 3), and should be assessed empirically for each type of spectrophotometer used.

For the same internal dimensions, plastic cuvettes are generally smaller outside than glass cells. In addition, semi-micro plastic

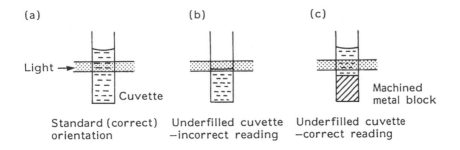

FIGURE 5.2: *Positioning the cuvette in the light beam. (a) Standard (correct) orientation. (b) Underfilled cuvette – incorrect reading. (c) Underfilled cuvette – correct reading.*

cuvettes (i.d. 1 cm × 0.4 cm) from some manufacturers do not have a square cross section. Thus there is a significant possibility of misalignment which is manifest, for example, in repeated measurements varying significantly when a single sample is removed and replaced in the instrument. If this is a problem, it may be overcome by using spring fittings to hold the cuvette in place or, in extreme instances, by changing the cuvette supplier.

A more insidious problem arises if the light beam is incorrectly positioned in relation to the vertical axis of the sample cuvette. If the cuvette is inadequately filled, the liquid level may be below the light path (widely varying readings on similar samples) (*Figure 5.2b*). Again, this will vary with the particular instrument – the position of the light source and how it is collimated/focused being the relevant factors. It is always advisable empirically to establish the minimum volume acceptable in any type of cuvette used. This can be done by measuring the apparent absorbance of a relatively small volume of sample, then increasing the volume in small steps, and repeatedly measuring absorbance until a stable reading is obtained.

It is, however, possible to exploit this behavior for minimizing sample volume, when sample size is experimentally limited. If, for example, the light beam passes through the uppermost regions of the cuvette, the base of the cuvette can be raised (often by as much as 1 cm) with no effect on the absorbance observed (*Figure 5.2c*). This means that the overall sample volume can be reduced with no change in position of the meniscus. Again, the extent to which this is allowable must be determined empirically for the instrument in question, but subsequently a machined aluminum block of suitable size can be left permanently in the cell holder.

5.2.4 Cleanliness

It is obvious that cuvettes should be kept clean – they should not contain, nor bear on their surface, any absorbing material. In addition, they should not contain compounds which might interfere with the absorbance of the material under study.

Plastic cuvettes are supplied clean and dry, and are normally disposable after one, or at the most, a few, measurements. It is thus normally sufficient to ensure they are clean on the outside (free from fingerprints and liquid droplets) by wiping with tissue before measurement.

Glass and silica cuvettes also need to be cleaned internally. For studies with routine, aqueous, biological samples, the cuvette should be emptied immediately after use, rinsed with water several times and then rinsed with ethanol or acetone (to denature traces of enzymes which may adhere to the surface). Traces of solvent should then be removed by drying internally and externally with, for example, a domestic hair dryer, before using again.

If the inside faces remain contaminated after this treatment – or the cuvettes are left for some time containing proteinaceous material which precipitates on the walls – cleaning may be effected by soaking the cuvette in a bath of 50% nitric acid overnight, followed by washing in water. Chromic acid should be avoided since it leaves a residue of metal ions which are difficult to eliminate completely – cuvettes which have been treated with chromic acid need to be washed with solutions of EDTA, to complex these ions, before use. Similarly, detergents may be effective cleansing agents, but the cuvettes should be well washed with water and ethanol after use, to remove all traces of detergent.

5.3 Preparing the sample

5.3.1 Medium

The sample should be dissolved or suspended in a nonabsorbing buffer. For measurements in the visible region, this presents no problem. However, some of the more complex biological buffers (HEPES, MOPS), developed by Good and co-workers, may absorb significantly below 230 nm. Phosphate buffers are convenient down to

190 nm. Commonly used buffer additives, such as ATP (ε_{260} =15 700 M^{-1} cm^{-1}) and dithiothreitol may also interfere with absorbance measurements in the ultraviolet.

5.3.2 Uniformity

The sample should be thoroughly mixed before measurement. This may be achieved by pre-mixing all components in a tube, using a vortex mixer. However, if the sample is prepared by repeated additions to the cuvette itself (e.g. by adding a concentrated stock to buffer in the cuvette), it must be mixed by stirring manually (*Figure 5.3*) or by covering the top of the cuvette with an inert film (e.g. Parafilm™) and inverting. Do not shake the cuvette vigorously, as this may cause foaming and/or trap air bubbles.

5.3.3 Clarity

If the sample is soluble, check that, after mixing, it produces an optically clear solution. This can be done simply by eye, although it is surprising how often this simple precaution is omitted. If the sample is cloudy, remove insoluble material by centrifugation (3000 *g*, 10 min is normally sufficient).

5.3.4 Stirring

If the sample forms a suspension (e.g. cells, membrane fragments, liposomes etc.), this too should be uniform (no large aggregates

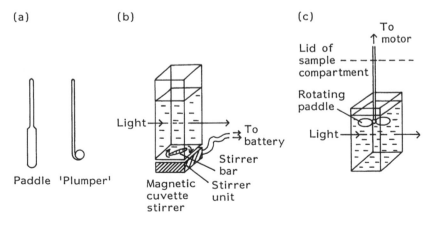

FIGURE 5.3: *Methods of stirring the cuvette contents within the instrument. (a) Manual devices. (b, c) Electrical devices.*

visible). If it is not, it should be homogenized using, for example, a Teflon™–glass hand homogenizer, or by sonication in a probe sonicator (e.g. MSE Soniprep).

Again, the sample should be thoroughly mixed before starting the measurement. However, over prolonged time periods, a suspension may settle and so repeated mixing may be required. Manual methods as discussed above can be used if measurement is intermittent. Otherwise, it may be necessary to stir continuously using, for example, a small magnetic bar inside the cuvette below the light beam (see *Figure 5.2*). This is driven by a small stirring block (conveniently 1 cm × 1 cm base × 0.2–0.5 cm high) which fits inside the sample holder (available from Temptron Electronics (*Figure 5.3b*). Less conveniently, an overhead stirring device may be present in the cuvette above the light beam.

5.3.5 Temperature

The solution should be brought to the temperature of measurement before it is added to the cuvette. Addition of an ice-cold solution to a cuvette at room temperature may lead to condensation on the outside cuvette walls, and this in turn produces spurious (and varying) absorption measurements. It is convenient to keep any diluent (buffer) at the measuring temperature during the period of the experiment, although this may not be possible with more labile components of the solution (such as enzymes).

6 Measuring an Absorption Spectrum

6.1 Introduction

The absorption spectrum of a compound, or a solution, is its absorbance as a function of wavelength. It is normally displayed as a plot of absorbance (A) on the ordinate against wavelength (λ) on the abscissa. Chromophores which are chemically distinct show different spectra, and thus an absorption spectrum can be used to identify chromophores present in a sample (see Chapter 2). We are concerned here with the absorption of visible and/or ultraviolet light, and thus measurements are made within the range 200–1000 nm.

The wavelength illuminating the sample is normally altered by varying the alignment of the monochromator (see Chapter 3). In modern instruments, this is achieved by rotating the grating so that different wavelengths fall on to its exit slit. This can be carried out manually using a rotating knob but, more conveniently, the grating is driven by a motor. In this case, the instrument is termed a scanning spectrophotometer. For spectra which need to be recorded very rapidly, this physical movement of the grating can limit the time scale available. In this case, simultaneous recording at a variety of wavelengths can be made using a diode array detector (see Chapter 4).

6.2 Calibrating the instrument

For accurate measurements of spectra, as for any absorbance measurement, it is essential that the spectrophotometer is correctly calibrated. This involves:

(i) ensuring that the wavelength reading on the instrument corresponds to the wavelength passing through the sample;

(ii) ensuring that 0% transmission ('dark current') and 100% transmission values are set at all wavelengths under study.

With modern instruments, these calibrations are usually performed automatically when the machine is switched on. Manual methods have been described in Chapter 5.

6.3 Selecting a sample

The sample, in uniform solution or suspension, should be prepared in a suitable cuvette as described in Chapter 5. As far as possible, neither the suspension medium nor cuvette should absorb light over the region to be measured. Rather than wasting valuable sample, it is always advisable to measure the 'absorption spectrum' of the cuvette containing all components of the sample apart from the compound under study. If this baseline measurement departs significantly from $A = 0$ at any particular range of wavelength, the buffer or cuvette material should be changed.

The concentration of sample should ideally be chosen so that its absorbance at any of the wavelengths studied is less than 2 (preferably $A < 1$). It is difficult to measure the position of a very high absorbance peak precisely, and the presence of such a peak (in particular the range of wavelengths where $A > 1$) may obscure other features of the spectrum. Samples with very high absorbance values should therefore be diluted, or measured in cuvettes with a decreased pathlength, so that the measured absorbance falls into a convenient range.

6.4 Choice of operating conditions

To measure an absorption spectrum, we need to set:

(i) the range of wavelengths to be scanned,

(ii) the spectral bandwidth of the illuminating beam (slitwidth),

(iii) the speed of scanning, and

(iv) the temperature of measurement.

6.4.1 Wavelength range

There are relatively few chromophores in biological systems (see Chapter 2), and their optical properties are well defined. For example, heme groups in proteins typically absorb between 400 and 650 nm, aromatic groups between 200 and 300 nm, etc. If the chromophore is known, therefore, a suitable scan range can easily be chosen.

If the chromophore is not yet known, it may be convenient initially to scan rapidly over a range of wavelengths (say 350–900 nm if the sample is colored, 200–350 nm if not) to identify absorption peaks. The width of scan is then reduced to cover the region of the absorption peaks, and scanning carried out more slowly. With today's digital collection methods, a narrower scan range allows more data points per nanometer and thus higher resolution. A slower scan rate also increases resolution (see below).

6.4.2 Spectral bandwidth

The light falling on the sample consists of a narrow band of wavelengths selected by the exit slit of the monochromator. This range of wavelengths defines the spectral bandwidth of the illuminating beam, which in turn affects the resolution of the spectrum – as discussed in Section 4.1.

In most present-day instruments, the slits are set automatically and are inaccessible to the user. Typical spectral bandwidths are less than 2 nm which are adequate for measuring most biochemical spectra without significant distortion. In older instruments, for highest resolution, the slitwidth should be set so that the spectral bandwidth is less than 15% of the natural bandwidth of the narrowest peak being measured – always assuming that this allows adequate light to reach the detector (i.e. the signal/noise ratio is still acceptable). If the relationship between slitwidth and spectral bandwidth is unknown (see instrument handbook), the slitwidth should be reduced stepwise until the signal/noise ratio on the spectrum becomes unacceptable, and a slightly higher value used for the measurements. Note, however, that this 'empirical slitwidth' will depend on concentration of the material under study, as this also affects the fraction of incident light reaching the detector, and hence the signal/noise ratio.

6.4.3 Scanning speed

For spectrophotometers with conventional optics, scanning is achieved by moving the monochromator grating stepwise, in small increments, using a stepping motor. At each position, the detector integrates the light arriving during the short 'dwell time' before the grating moves on. This signal is then passed to the recorder, which notes the signal at each (median) wavelength and this records the absorption spectrum. The rate of scanning should be such that the recorder can keep pace with the wavelength variation, otherwise the spectrum will be distorted (*Figure 6.1*).

The choice of scan speed will thus be influenced by three instrumental factors:

(i)   the rate at which the monochromator can physically be moved (see above);

(ii) the sensitivity of the detector and the lamp intensity – enough light must fall on to the detector during the dwell time to produce a good signal/noise ratio; and

(iii) the response time of the recorder.

In modern instruments, detector sensitivity is rarely a factor limiting the rate of data collection. It is, in any case, rarely under control of the operator. If sensitivity is a problem, because measurements need to be made on solutions of very high absorbance, a change of instrument would be indicated. The new instrument should have:

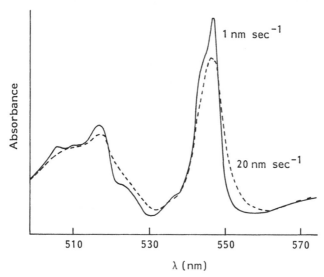

FIGURE 6.1:   *Effect of scan rate on resolution. The spectrum is that of cytochrome c (reduced–oxidized) at 77 K. Note the broadening and loss of resolution of the peaks at the high scanning rate (cf. Figure 4.3). Data from Poole and Bashford (1987)* Spectrophotometry and Spectrofluorimetry *(D.A. Harris and C.L. Bashford, eds), pp. 23–48. IRL Press, Oxford.*

(i)   a more intense light source (so that more light falls on the detector); and/or

(ii)   a more sensitive detector.

An instrument using a photon counting detector (rather than the conventional photomultiplier or photodiode) may be considered if high sensitivity is required.

If data is being recorded by a pen recorder (flat bed x–t or x–y recorder), the inertia of the pen itself can limit the available scanning rate. For an accurate record, the movement of the pen across the chart must at least match the speed of the stepping motor. Again, the instrument manual will often suggest suitably matched scan speeds/chart rates. However, it is often wise to check this empirically by running the same spectrum at the recommended and at half the recommended speeds, to confirm that the spectral resolution is unchanged. Some instruments allow the user to select a response time – and/or change the 'damping' on the chart recorder – and this allows limited control over available scan speeds, again by empirically balancing distortion (at high damping) against noise (at low damping).

With the advent of electronic data capture (either within the instrument itself, or by transmission to a stand-alone computer), recorder rates need no longer limit scanning rates. An analog signal (from the photomultiplier) can be converted, instantaneously, to a digital signal and rapidly recorded on a magnetic disc. None the less, on conventional instruments, the speed of response is rarely increased by more than one order of magnitude (perhaps down to 10 msec response time) by computer data capture. In the first place, depending on the attendant software, the speed with which data is passed between the digital output and the storage medium may not be extremely fast (see manual). Secondly, even if the system is optimized for rapid data transfer, speed may now depend on detector sensitivity, that is, the integration time at each position of the stepping motor. To decrease response time significantly, therefore, a more intense light source will ultimately be required.

6.4.4 Temperature of measurement

Absorbance spectra, unlike fluorescence spectra, are hardly affected by temperature between 0–4°C. Thus it is normally convenient to measure absorption spectra at ambient temperature (20–25°C), particularly when such spectra will be used to identify chromophores present.

To follow changes in absorbance with time, particularly where these are due to chemical reactions – controlling the temperature of the cuvette becomes important, since temperature will affect the rate of change observed. Most spectrophotometers can be equipped with thermostatically regulated cuvette holders so that the temperature of the cuvettes can be controlled between −40°C and +60°C. Control can be achieved by circulating water (or a water–ethylene glycol mixture) from a thermostatically controlled bath through a water jacket surrounding the cuvettes. A much more rapid control (albeit much more expensive) is achieved by a heating system employing the Peltier effect.

Below −40°C, a more complex 'low temperature cuvette holder' is required. Liquid nitrogen temperatures (about −190°C) are convenient for fixed temperature work, although variable temperature devices are available. The use of such low temperatures slows down the chemical processes, and allows spectroscopic studies on unstable intermediates in reactions; an increase in spectral resolution also occurs (see Section 6.8).

A typical (commercially available) low temperature accessory is shown in *Figure 6.2*. The cuvettes (typically perspex) are held in the light beam within a (part silvered) Dewar vessel, which is filled with

(a)   (b)

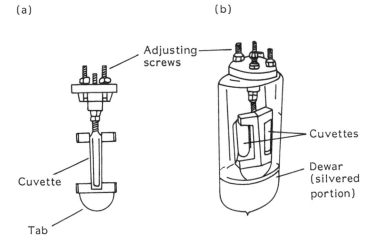

FIGURE 6.2: *Device for measuring spectra at low temperature. The cuvettes are held in a holder (a), which is then immersed in liquid nitrogen in a part-silvered Dewar flask (b) which is mounted in the sample compartment of the spectrophotometer. (Note part (b) is not drawn to scale). Reproduced from Gottschalk (ed.) (1985)* Methods Microbiol.*, **18**, p. 285, with permission from Academic Press.*

liquid nitrogen up to the level of the cuvette base. The sample itself is dissolved/suspended in a buffer (preferably phosphate, whose pK_a varies little with temperature) and frozen with glycerol or sucrose present to promote the formation of a uniform polycrystalline phase.

6.5 Determining a baseline

To measure the absorption spectrum of a compound X, we must ensure that any variation in signal observed (as the absorbance of a solution of X is measured across various wavelengths) is due to the absorption of light by X itself. Two factors act to frustrate this aim.

(i)   The intensity of lamp emission and the sensitivity of the detector will vary with wavelength, and thus the detector current produced will vary with wavelength, even if no absorbing material is present in the light beam.

(ii)   Other components of the system apart from X – for example, the cuvette, the buffer, added components Y and Z in the solution – may also absorb light in the region under study.

In most modern instruments (provided they are serviced regularly), the variations in signal due to lamp and detector response are compensated for electronically (see Section 3.6), and thus a spectrum taken with no sample in the light beam (electronic baseline) should be flat. If this is markedly not the case, or noise in the baseline is high, consult the instrumental manual.

Absorbance due to other components of the system can be minimized by appropriate choice of materials – silica cuvettes below $\lambda = 300$ nm, low absorbing buffers, and omission of additives such as ATP which have strong absorbance bands. In solutions of biological materials, this is rarely a problem in the visible range, but some baseline absorbance may be unavoidable in the ultraviolet. In suspensions, on the contrary, turbidity will generate apparent baseline absorbance even in the visible range, and accurate absorbance measurements may require specialized techniques (see below).

Assuming precautions have been taken to minimize baseline absorbance, the absorption spectrum is obtained by:

(i)   measuring the spectrum of all components of the system without X present (baseline);

(ii)   measuring the spectrum of all components of the system with X present;

(iii)   subtracting spectrum (a) from spectrum (b).

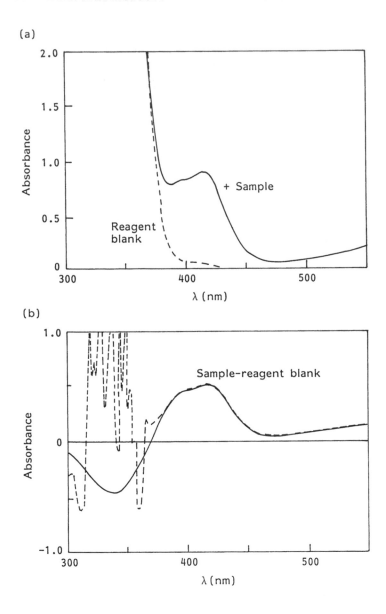

FIGURE 6.3: (a) Single beam spectrum of ABTS (0.5 mM oxidized by glucose (20μM) (—), or unreacted (···). (b) Difference spectrum of the two curves in (a), as calculated by the instrument (···). Note the high noise level generated by the substraction of two large numbers. The solid line represents the true difference spectrum (measured at 0.1 mM ABTS).

This procedure is shown diagrammatically in *Figure 6.3*. Note that, where the baseline absorbance is high, the net signal is 'noisy', since it represents the small difference between two large numbers (high per cent error).

In fact, baseline subtraction is rarely performed manually. The actual method used depends on the type of instrument available. For single beam instruments, spectra of 'blank' (baseline) and 'sample' solutions are taken sequentially, as discussed above, but the baseline measurement is typically stored digitally within the instrument, which then subtracts it from subsequent spectra. Even relatively cheap instruments today have sufficient built-in computing power to correct spectra in this way.

In double beam instruments, on the contrary, a spectrum can be adjusted for baseline absorbance during a single measurement, in which the sample is placed in the 'measuring beam' and the blank cell in the reference beam. The resulting signal directly records the difference between the sample and the baseline and thus the spectrum of the single component absent in the blank is obtained.

While it is, therefore, relatively simple to obtain corrected spectra on modern instruments, as in all automated procedures there may be pitfalls for the unwary. Firstly, it is essential that the only difference between the blank and sample cells is the presence of X. Since, particularly in the double beam instrument, the sample and blank solutions are likely to be present in different cuvettes, it is important that the absorbance of the two cuvettes is identical over the wavelengths studied, that is, that the cuvettes are 'matched'. This is a particular problem for measurements in the far ultraviolet region, where different batches of cuvettes may vary in absorbance. Secondly, the automatic subtraction of blank from sample spectrum may obscure a high baseline absorbance, and thus a high uncertainty, on the absorbance values obtained over certain regions of the spectrum. This is frequently observed as noise on the resultant difference spectrum (*Figure 6.3b*). Therefore, it is always advisable to inspect the baseline spectrum directly, rather than simply allow the instrument to use it, in order to see whether such problems are occurring. In the case of a double beam instrument, blank versus air, and blank versus blank spectra should be measured.

6.6 Interpreting spectra

6.6.1 Identification of compounds

Each absorbing compound will have its own unique spectrum. The spectrum obtained for an unknown compound can thus be compared with spectra from the literature for identification. The particular features of the spectrum that are useful in identifying compounds are (i) the position of the absorption peaks (λ_{max} values) and (ii) the pairwise ratios of peak heights ($A_{\lambda 1}/A_{\lambda 2}$). Both of these are independent of concentration (see *Figure 8.4*).

The position of peaks can be determined manually, by dropping a vertical line from a maximum in the absorption spectrum to the wavelength axis (abscissa). If a computer display is available, this process can be aided by scrolling a cursor across the curve until a maximum is reached. Alternatively, software is widely available to identify maxima in absorption spectra by differentiation, a maximum occurring when the first derivative ($dA/d\lambda$) = 0. The ratio of peak heights between peaks can also be determined manually or by using widely available software.

The measured values can then be compared with relevant values from the literature. For any given compound, the values will correspond only if (i) the instrument is correctly calibrated (Chapter 5); (ii) the conditions of measurement (notably the pH in biological systems) correspond to the literature conditions; and (iii) the scan speed and response time of the instrument are suitable. This last point is discussed in Section 6.4.3, and means that, for example, while scanning from high to low wavelength, if the scan speed is too high (relative to the response time), peaks will appear broadened and shifted to lower wavelength. To eliminate this possibility, the spectrum should be recorded a second time, at a slower scan speed, to confirm that no change is observed.

6.6.2 Measuring concentration

The concentration of a pure compound in solution can be determined from its absorbance provided its molar absorptivity at that wavelength is known. Molar absorptivities are normally tabulated for absorption at peaks within a spectrum i.e. λ in the parameter ε_λ refers to an absorption peak. ε_λ is the absorption of a 1 mol l^{-1} solution in a 1 cm light path at λ_{max} . Thus

$$c = A_\lambda / \varepsilon_\lambda \ \text{mol} \ l^{-1}$$

This is a widely used method of determining concentrations in biochemistry.

Conversely, if the concentration of a compound is known, A_λ can be used to determine molar absorptivity. Note, however, that determining the concentration of biological materials is not always straightforward. A simple measurement of weight – of a protein to be dissolved, for example – requires careful interpretation, as the weighed material will necessarily contain not only the polypeptide chain but also counterions, bound water and, frequently, additional salt. In addition, molecular weights are rarely known precisely for macromolecules. In summary, determination of molar absorptivities of biomolecules requires not only precise spectrophotometry but also detailed characterization of the material under study.

6.6.3 Characterizing mixtures

The absorbance of a mixture of compounds, at any wavelength, is the sum of the individual absorbances of each compound alone (providing the compounds do not interact with each other). As a corollary, the spectrum of a mixture of compounds is the sum of the individual spectra.

If the absorption peaks of the compounds involved can be resolved, compounds can be identified in mixtures by λ_{max} values as for pure compounds. Individual cytochromes, for example, can be identified in mitochondrial membranes that contain a number of different cytochromes (*Figure 6.4*). Even when complete resolution is not possible, if the spectra of the individual pure compounds is known, it is possible to simulate a measured spectrum by (computer) addition of different levels of the individual spectra, and thus confirm the presence (or absence) of particular components (*Figure 6.5*).

In principle, quantification of individual components in a mixture is performed as above, using ε_λ for each component. However, in a mixture the actual absorbance, A_λ, will be the sum of absorbances for each component at that wavelength. Thus for a two component mixture, two measurements are required ($A_{\lambda 1}$, $A_{\lambda 2}$) together with the molar absorptivity of each component at each wavelength required ($\varepsilon^a_{\lambda 1}$, $\varepsilon^a_{\lambda 2}$ ...) to solve the simultaneous equations for the concentrations c_a, c_b

$$A_{\lambda 1} = c_a \varepsilon^a_{\lambda 1} + c_b \varepsilon^b_{\lambda 1}$$

$$A_{\lambda 2} = c_a \varepsilon^a_{\lambda 2} + c_b \varepsilon^b_{\lambda 2}$$

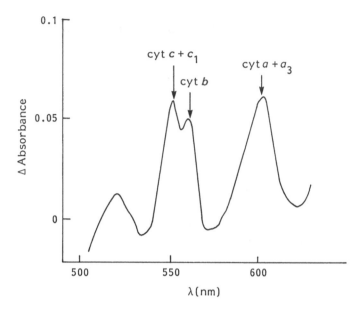

FIGURE 6.4: *Identification of cytochromes in mitochondrial membranes. The spectrum shown is the difference spectrum (reduced–oxidized) of ox-heart mitochondria in suspension. Data taken from Rickwood et al. (1987) Mitochondria: a Practical Approach (V.M. Darley-Usmar et al., eds), pp. 1–16, by permission of Oxford University Press.*

Thus c_a and c_b can be calculated from measurement of absorption at two wavelengths.

A classical example of this approach is the measurement of protein and DNA concentration in a mixture of both. Measurements are made at $\lambda = 260$ nm, where $\varepsilon_{1\%}^{NA} = 220$ cm^{-1}, $\varepsilon_{1\%}^{prot} = 3.77$ cm^{-1}) and 280 nm ($\varepsilon_{1\%}^{NA} = 108$ cm^{-1}, $\varepsilon_{1\%}^{prot} = 6.60$ cm^{-1}) and the concentrations of each in the mixture can be determined.

6.7 Working without oxygen

One important application of spectrophotometry in biochemistry has been to study redox centers in proteins. These frequently contain chromophores, such as the heme group in hemoglobin and the cytochromes, which are convenient for such investigations. However,

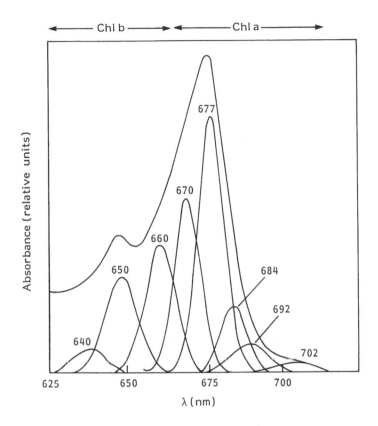

FIGURE 6.5: *The spectrum of spinach chloroplasts (upper line) deconvoluted into peaks from individual classes of chlorophyll molecules (lower curves). The spectrum was measured at 77 K. Redrawn from French et al. (1972) Plant Physiol.* **49,** *p. 429, with permission from the American Society of Plant Physiologists.*

by their nature, many of these centers may be oxidized by oxygen from the air. Thus to study such chromophores in a variety of redox states (as in a redox titration), it may be necessary to carry out measurements in an oxygen-free environment.

Effectively, such measurements require a cuvette isolated from the atmosphere. The cuvette is sealed with an airtight stopper, through which passes a gas inlet and gas outlet port (needle). Solutions are degassed by bubbling with argon, and added to the lower part of the cuvette, also under argon, through a separate injection port. The cuvette can then be placed in the light path. If necessary, the solution may be stirred by an internal stirrer (see *Figure 5.3*), and further additions made through the injection port.

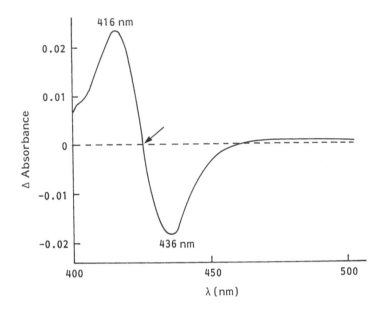

FIGURE 6.6:   *Shift of γ band of cytochrome o on formation of the carbon monoxide complex. The spectrum is that of (CO derivative-reduced)* E. coli *membranes at 77 K. Only cyto contains a free heme coordination site capable of binding O_2 (or CO). A shift of the peak (436 nm → 416 nm) is seen by a loss of absorption at 436 nm and a gain at 418 nm. The arrow indicates an isosbestic point. Data from Poole and Bashford (1987)* Spectrophotometry and Spectrofluorimetry: a Practical Approach *(D.A. Harris and C.L. Bashford, eds), pp. 23–48. IRL Press, Oxford.*

Besides excluding oxygen, such cuvettes are useful for maintaining samples under other atmospheres. For example, carbon monoxide may be bubbled into the solution to form the C≡O adduct of the redox center under study. The C≡O adduct of hemes shows a characteristic spectral band, (see, for example, *Figure 6.6*). Since only heme groups with an open co-ordination site (hemoglobin, cytochrome oxidase for example) can bind C≡O, reaction with this compound has proved useful in identifying heme groups with an open co-ordination site in a complex mixture of other redox centers within, say, bacterial cell membranes.

6.8 How to improve resolution

The resolution of a spectrophotometer is its ability to distinguish between two closely spaced absorption bands. As noted above, this is maximized by (i) choosing a sufficiently narrow slitwidth (spectral bandwidth) so that the observed peak width is no broader than its natural bandwidth (Section 6.4.2) and (ii) choosing an appropriate scan speed and response time (Section 6.4.3). However, in systems whose spectra are complex because (i) multiple absorbing components are present and/or (ii) significant light scattering occurs, additional measures may be needed to resolve individual components.

6.8.1 Difference spectra

The most widely used technique for enhancing resolution is to measure a difference spectrum. In principle, the method is identical to that described above for subtracting a baseline; the spectrum of a reference is subtracted from that of the sample under study either in a single measurement (using a dual beam instrument) or sequentially, by storing the reference data in the instrument prior to measuring the sample (in a single beam instrument). The difference in method lies in the composition of the two solutions used: to measure a difference spectrum both sample and reference contain the chromophore under study, but in the sample, the chromophore has been chemically modified. Thus, for example, a difference spectrum of reduced minus oxidized cytochrome c is shown in *Figure 6.7a*, in comparison with the spectrum of each individually (see Chapter 1, *Figure 1.8*). The difference spectrum clearly shows that, on reduction, a peak appears at 550 nm, while absorbance falls at about 535 nm. Difference spectra (light minus dark) of chloroplasts reveal both cytochrome oxidation and the photochemical reaction centers, which are bleached (at 682 and 700 nm) on illumination (*Figure 6.7b*). Similar observations can be made, for example, during the reaction of a protein with a chromogenic reagent (*Figure 6.8*).

The increase in resolution resulting from this approach is clearly seen in *Figure 6.9*, where the spectrum of a highly turbid yeast suspension is shown. The 'absolute' spectra of the suspension, in both reduced or oxidized state, are dominated by light scattering from the solution and, as a result, appear featureless and very similar. However, in the reduced minus oxidized difference spectrum, a number of peaks are resolved, corresponding to various cytochromes within the cell (the peak at 550 nm representing cytochrome c, for example). The levels and redox potentials of these individual components *in vivo* can therefore be studied.

Referring back to *Figures 6.6–6.8*, features of difference spectra can be seen to include positive peaks (higher absorbance under sample conditions: appearance of a product), and negative peaks (higher

(a)

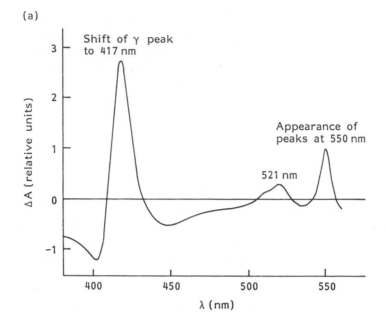

(b)

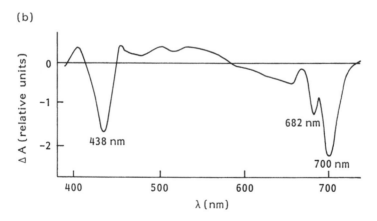

FIGURE 6.7: *Difference spectra: (a) (reduced–oxidized) cytochrome c showing the shift of the γ band (411 → 417 nm) and appearance of sharp α and β bands on reduction (see Figure 1.8). (b) Chloroplasts (illuminated-dark) showing cytochrome oxidation (decrease in absorption at 438 nm) and disappearance (bleaching) of chlorophyll species absorbing at 682 and 700 nm (see Figure 6.5). Reproduced from Witt (1979) Biochim. Biophys. Acta, **505,** p. 361, with permission from Elsevier Science Publishers.*

absorbance under reference conditions: disappearance of a reactant). There are also points where absorbance is identical in both sample and reference conditions, namely, $\Delta A = 0$ over the change studied and

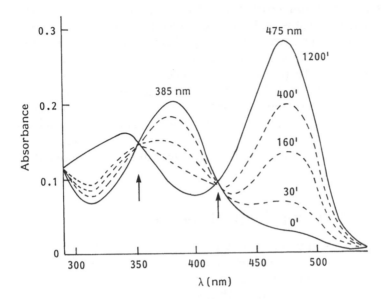

FIGURE 6.8: *Conversion of (tyr-O-Nbf) derivative of mitochondrial ATPase* (λ_{max} = 385 nm) to lys-N-Nbf derivative (λ_{max} = 475 nm). The conversion occurs by intramolecular transfer at pH 9.0. Spectra were measured at the times indicated: the half time for the transfer was 150 min. Isosbestic points between the two forms are indicated by arrows. Data from Ferguson et al. (1975) Eur. J. Biochem., **54,** pp. 127–133.

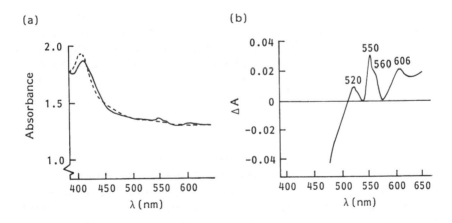

FIGURE 6.9: *Increase in resolution by difference spectroscopy.* **(a)** *Apparent absorbance of (turbid) yeast suspension native* (···) *or reduced with dithionite.* (—) **(b)** *Difference spectrum (reduced–oxidized) of same suspension. Note large difference in scale between* **(a)** *and* **(b)**. *Data from Jones and Poole (1985)* Methods Microbiol., **18,** p. 285.

the difference spectrum crosses the baseline. These are isosbestic points (iso, equal; sbestos, extinguished) (see Section 1.4). Isosbestic points occur when there are only two species involved in the change, that is, a single reactant produces a single product. (Note that the presence of an isosbestic point does not prove that only two species are involved: there may be intermediates with no absorbance in the region studied.) If no isosbestic point is observed during a process, at least one additional intermediate must be participating. Isosbestic points are useful in quantitating total levels of both components present, and also in setting up dual wavelength spectrophotometers (see the following section).

6.8.2 Dual wavelength spectroscopy

Difference spectroscopy is clearly a powerful technique for increasing spectral resolution. This is enhanced by the advantages of dual beam spectroscopy for decreasing instrumental noise (Chapter 4). However, one drawback of the technique is that separate reference and sample cuvettes must be prepared, for example, that comparison is between two different solutions/suspensions.

This may be a problem if the two solutions vary in time over the course of the experiment, notably in turbid (highly light scattering) suspensions. For example, particles in a suspension may gradually settle out over a period of time, changing the light scattering properties of sample and reference in an ill-defined, unmatched manner. Alternatively, the treatment accorded to the 'sample' may cause the suspended particles to swell or shrink, again changing their ability to scatter light. This is particularly noticeable in mitochondrial suspensions, where reduction/oxidation may affect mitochondrial volume very significantly. In both these examples, the apparent 'absorbance' of the sample will differ from that of the reference as a result of changes in light scattering in addition to any change in the chromophore studied.

These problems may be overcome by using dual wavelength spectrophotometry, in which the sample beam and reference beam pass through the same cuvette. This is achieved by a rotating mirror (chopper) which alternates the two beams through one sample (Chapter 4). Typically the reference wavelength is kept constant, while the sample beam is scanned across a range of wavelengths. Ideally, the reference wavelength is set at an isosbestic point of the change under study – thus any changes in the reference beam are due to changes in light scattering by the sample and are subtracted automatically from the signal due to the sample beam. Since the variation of light scattered with wavelength is relatively small and fairly linear, the subtraction of a reference beam at one wavelength from a sample signal over a range of nearby wavelengths is sufficient to remove much of the interference due to light scattering, and hence to increase resolution in the resultant spectrum (*Figure 6.10*).

6.8.3 Low temperature spectroscopy

Techniques for low temperature spectroscopy have been described above (Section 6.4), as methods for observing short lived species in chemical reactions. The lifetime of these intermediates is increased at low temperatures and reaction rates are slowed down, and they can thus be observed more easily.

However, in frozen solutions (particularly between −40°C and −196°C), spectral resolution also improves markedly as absorption bands become narrower (due to vibrational effects on molecular energy levels) and enhanced (due to multiple internal reflections

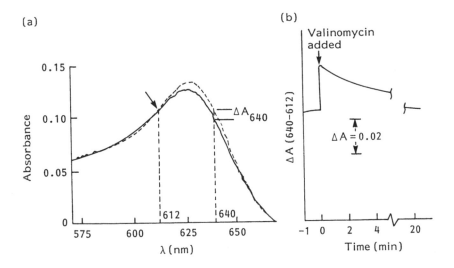

FIGURE 6.10: *Use of dual wavelength spectroscopy. (a) The spectrum of oxonol V in liposomes, (—) in the absence of membrane potential. (···) in which the membrane potential has been induced by the addition of valinomycin (in the presence of KCl). A small change in absorption is observed (see Section 2.3). The arrow indicates an isosbestic point. (b) The time course of decay of this potential as measured by $\Delta A_{(640-612)}$, the isosbestic point being used as a reference wavelength to eliminate noise due to light scattering effects etc. Reproduced from Fleischer and Packer (1979) Methods Enzymol., **55**, p. 579, with permission from Academic Press.*

within the polycrystalline phase), becoming as much as 20–30 times more intense. This is demonstrated in the spectra of cytochrome c, at 25°C and −196°C, shown in *Figure 6.11*. Thus, provided the system under study is amenable to freezing (and does not become irreversibly damaged in the process), resolution of spectra can be significantly enhanced by work at very low temperatures.

6.8.4 Derivative spectra

Three approaches to increasing spectral resolution by improving the quality of the data collected, that is, by improving the instrument or changing conditions of measurement, have been discussed. Of the three, only difference spectroscopy (either single or double beam modes) is widely available on commercial instruments. Dual wavelength and low temperature spectroscopy are less common and usually require more expensive, specialist instruments.

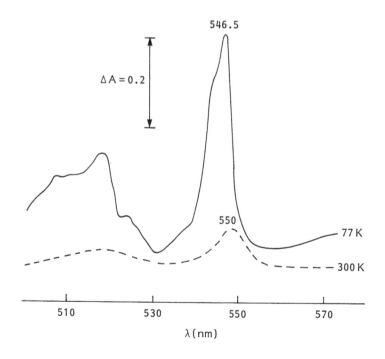

FIGURE 6.11: *Spectra (reduced–oxidized) of cytochrome c at 77 K and at room temperature. Note change in intensity of peaks observed, increased resolution and slight blue shift of peaks at the lower temperature. Data from Poole and Bashford (1987)* Spectrophotometry and Spectrofluorimetry: a Practical Approach *(D.A. Harris and C.L. Bashford, eds), pp. 23–48. IRL Press, Oxford.*

An alternative approach to maximizing resolution is to modify analysis of the data obtained. We have already seen how differentiating a spectrum can aid in determining the precise position of absorption peaks. Determining higher derivatives ('derivative spectroscopy') may also prove useful, particularly in resolving partly overlapping bands.

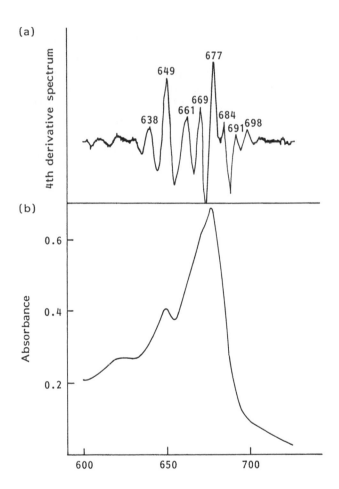

FIGURE 6.12: *Identification of spectral components by derivative spectroscopy. (a) 4th derivative (differentiation using 4 nm intervals). (b) Absorption spectrum of choroplasts at 77 K. Note the identification of components corresponding to those in* Figure 6.8b. *Adapted from Butler and Hopkins (1970)* Photochem. Photobiol., **12,** *p. 453, with permission from the American Society for Photobiology.*

In principle, the first and third derivatives produce curves which cross the baseline at the positions of the original peaks and troughs, while the second derivative gives a trough where, in the undifferentiated signal, there was a peak. Perhaps the easiest to interpret visually is the fourth derivative where peaks correspond to peak positions (λ_{max} values) in the original spectrum. *Figure 6.12* shows the ability of a fourth derivative spectrum to increase resolution of chlorophyll species in chloroplast membranes (cf. *Figure 6.8b*)

The calculation of derivative spectra is conveniently carried out by a microcomputer linked to (or an integral part of) the spectrophotometer, using software supplied by the manufacturer. Problems include effects of noise in the original spectrum (which may be overcome by summation of several spectra and/or computer smoothing of the spectra), and the possibility of 'losing' broad peaks, or generating 'spurious' peaks during the derivatization. None the less, providing its limitations are borne in mind, derivative spectroscopy can be a useful aid to interpreting spectrophotometric data.

7 Measuring a Fluorescence Emission Spectrum

7.1 Types of fluorescence spectrum

There are, in principle, two types of fluorescence spectrum, the excitation spectrum and the emission spectrum. An excitation spectrum is measured by fixing a convenient wavelength at which to measure fluorescence and then following fluorescence intensity as the wavelength of illumination is varied. Since only absorbed light can be re-emitted, the shape of an excitation spectrum should be similar to that of an absorption spectrum measured as described in Chapter 6, and hence it gives little additional information. An excitation spectrum may be more easily measured than an absorption spectrum, as fluorescence measurements are inherently more sensitive than absorbance measurements (see Section 2.5), and so such measurements are useful if amounts of sample are small.

The drawback to this approach is that, in many fluorometers, no correction is made for variation in lamp intensity with wavelength, and hence the relative peak intensities observed in an excitation spectrum will depend on the nature of the lamp as well as on the absorbance of the sample. Spectra 'corrected' for this effect can be obtained most easily by using a split beam fluorometer where a fraction of the exciting beam is split off and measured independently (see Chapter 4), but such instruments may be beyond the resources of all but specialized laboratories.

More useful information can be obtained by measuring an emission spectrum. In this case, the wavelength of illumination is fixed (commonly at, or near, an absorption peak of the fluorophore) and the emitted light is sampled over a range of wavelengths. Sampling over a range of wavelengths can be achieved by driving the grating of a monochromator (the 'emission' monochromator, see *Figure 4.9*) or by dispersing the emitted light across a diode array detector.

The procedure for measuring a fluorescence spectrum is in many respects similar to that described for measuring an absorption spectrum (Chapter 6). In the following discussion, the differences between the procedures will be highlighted.

7.2 Calibrating the fluorometer

(i)   The user should ensure that the wavelength readings on the instrument correspond to the actual excitation/emission wavelengths used. Fluorescent standards (frequently plastic blocks doped with a fluorescent compound for example tetraphenyl butadiene) are available for calibration measurements.

(ii) The user should ensure that the photomultiplier gives no signal in the absence of a fluorescent sample and/or when the exciting light is blocked with a shutter.

7.3 Preparing the sample

For right angle fluorescence measurements, the sample should be placed in a cuvette with four clear walls, rather than, as for absorbance measurements, only one pair of opposite faces clear. If the excitation wavelength is less than 340 nm, the cuvette should be silica or, for the near ultraviolet, an appropriate plastic (Section 3.4). For measurements of front face fluorescence, it is possible to use the same cuvettes as those used for absorbance measurements. Neither buffer nor cuvette should fluoresce over the range of measurement, and this may be determined from a 'baseline' scan in the absence of fluorophore.

In fact, baseline fluorescence is frequently a problem. One source of such fluorescence might be the water used for making solutions. Water purified on deionizing columns or stored in plastic bottles may acquire fluorophores which are present as trace components of the plastics with which it has been in contact. These can be conveniently removed by filtering the water through charcoal and storing it in acid-washed glass vessels. Occasionally, batches of commercial buffers may contain fluorescent impurities and in this case, alternative batches must be sought.

A more common source of interference in fluorescence measurements is light scattered from the sample. Fluorometers detect both Rayleigh scattering (at the same wavelengths as the excitation) and Raman scattering from the solvent (40–100 nm higher than excitation wavelength, Chapter 1). Scattering is, of course, highest in turbid solutions, where it may, in fact, swamp the right angle fluorescence signal. Under these conditions, useful fluorescence measurements can be made only using front face detection (Section 4.9). However, even in 'optically clear solutions' dust particles etc., passing through the light beam can cause large fluctuations (noise) in the measured fluorescence signal. All solutions must be filtered (0.2 μm or 0.45 μm pore size filter) before they are used in fluorescence measurements.

Another factor to be considered is the concentration of the sample. Over a range of concentrations, the emission from a fluorophore will be proportional to the concentration of fluorophore present. Deviations from linearity will occur:

(i)   if the molecules in solution aggregate or otherwise interact with each other to form a 'new' or modified chromophore;
(ii) if the absorbance of the solution is so high that the exciting light is absorbed within 1–2 mm of its entering the cuvette. In this case (the inner filter effect), no light will reach the centre of the sample, and thus fluorescence will not be apparent to the detector (see Chapter 4, *Figure 4.12*).

Except in particular situations (e.g. measurements within cells or bilayers), fluorescence should be measured in the linear region. As a check, dilutions of a fluorophore should be chosen so that the observed signal at any wavelength is doubled when the concentration used is doubled. As a rule of thumb, to avoid inner filter effects, the absorbance of a solution should be kept below 0.05 cm^{-1} at all wavelengths where fluorescence is to be measured.

7.4 Selecting the wavelengths and slitwidth for scanning

When measuring an emission spectrum, an excitation wavelength is chosen at, or close to, an absorption peak – conveniently a strong, high wavelength absorption peak of the particular chromophore. The emission wavelength is then scanned over a convenient range, largely determined by trial and error, as for absorption spectroscopy. However, since fluorescence will always occur at a wavelength higher than the absorbed radiation, from energetic considerations (see Chapter 1), the starting point for an emission spectrum is typically 10–20 nm higher than the excitation wavelength, with the detected wavelength subsequently rising.

The user must also set the slitwidth (spectral bandwidth) of both the exciting beam and the emission beam. In contrast to the situation with spectrophotometers, fluorometers normally require/allow the user to define the most appropriate slitwidths. As the emission spectrum is largely independent of illumination wavelength, it is convenient in most cases to select a wide excitation slit (say 20 nm bandwidth), to ensure that plenty of light reaches the sample. In contrast, a narrow emission slit should be chosen (2.5–5 nm bandwidth), for a good resolution of spectral bands (i.e. to ensure that the natural bandwidth is measured, and that the spectrum is not instrumentally broadened, see Section 4.1).

Occasionally it is not appropriate to select either (a) the wavelength corresponding to an absorption peak or (b) a wide slit for the excitation beam. Different cases are outlined below:

(i) If the absorbance of the sample solution is too high (and the inner filter effect predominates), this may be alleviated by selecting an alternative wavelength (where absorption is lower) for excitation.
(ii) If light scattering from the sample interferes with the fluorescence signal, the excitation slit should be narrowed and the excitation wavelength may be lowered.
(iii) If the sample is affected by photobleaching (see Section 7.6), the excitation slit should be narrowed.

7.5 Selecting the speed of scanning

For fluorescence measurements, as for absorbance measurement (see Chapter 6), wavelength scanning is accomplished by driving a monochromator with a stepping motor, and measuring the signal at each position of the monochromator. Scan speeds are thus limited by (i) the rates at which the monochromator is driven; (ii) the sensitivity of the detector and (iii) the response time of the recorder. Precisely the same considerations apply as discussed in Chapter 6. Empirically, if there is some doubt as to whether a scan speed is too fast, the scan should be repeated at half the speed. If the original scan speed was appropriate, no change in the spectrum should be observed when it is repeated.

Probably the most likely factor to limit scan speed in a fluorescence instrument is detector sensitivity; in fluorescence measurements the light levels reaching the detector are normally much lower than in absorbance measurements. Thus typically, fluorometers (i) have more intense light sources than spectrophotometers, and (ii) photon counting detectors are in more common use (see Chapter 3).

7.6 Photobleaching

It was noted in Chapter 1 that illumination of chromophores with visible or near ultraviolet light rarely results in damage to the chromophore. However, with intense light sources such as may be used in fluorometry, the fluorescence of a particular chromophore may be found to decrease over the time of measurement. This may be due to the ability of the intense light to induce photochemical reactions, where the excited molecule reacts with some other component of the system (and thus the concentration of chromophore declines). Rarely there may also be 'over excitation' of the chromophore. In this case, light is absorbed faster than the absorbed energy can be dissipated (as heat and light emission), in which case the number of ground state (absorbing) molecules decreases in time.

If fluorescence is observed to decrease with time, various precautions can be instigated. With mild cases of photobleaching, it may be sufficient (i) to decrease the time of illumination, by closing the excitation shutter when readings are not being made, and (ii) to stir the solution periodically, since only molecules in the light beam are bleached. In more severe cases, the intensity of illumination must be reduced, by narrowing the excitation slit (see Section 4.7), by exciting with a wavelength closer to the red end of the excitation spectrum, or by using a fluorometer (e.g. Perkin-Elmer LS series) with a pulsed light source to decrease the overall illumination period.

7.7 Determining a baseline

While it is common for solutions of biological interest to show no absorption over a range of wavelengths (in particular over the visible region), such solutions rarely show a zero baseline when a fluorescence spectrum is measured. Even if all contaminating fluorophores are absent, Rayleigh and Raman scattering will yield an irreducible minimum signal towards the low wavelength end of the scan (see Chapter 1, *Figure 1.11*). The detector signal, as the 'blank' cell (all components apart from the fluorophore) is scanned, provides the baseline, which must be subtracted from the sample signal itself.

In contrast to the situation with spectrophotometers, where double beam instruments are widely available, most commercial spectrofluorometers operate in the 'single beam' mode. (Note that, for a spectrofluorometer, measurement of a single sample requires both excitation and emission monochromators, so measurement in a 'difference' mode, sample minus reference, would require four monochromators!) Thus subtraction of the baseline from a sample measurement would typically be carried out manually or by using software to memorize and subtract the baseline, as described for 'single beam' operation of spectrophotometers (see Chapter 6).

There is a problem with this approach if addition of the chromophore itself increases light scattering by the solution, as may be the case if the chromophore is part of or bound to a macromolecule (*Figure 7.1*). If such effects occur, they can be eliminated only by carrying out appropriate control experiments and applying the relevant correction.

7.8 Determination of λ_{max} and peak intensity

The position of the peaks in a fluorescence emission spectrum can be determined as described for absorbance peaks – manually (by dropping a vertical on to the abscissa), on a visual display (using a cursor) or by computation (by determining the first derivative). Compounds can be characterized by the position of their fluorescence peak(s) under standard conditions.

More problematic is the expression of peak intensity. For absorption measurements, the intensity units (absorbance) are determined by directly comparing the amount of light impinging on the sample with that transmitted ($A = \log I_o/I$), and are thus independent of the instrument used. For fluorescence, measurements are made of only a fraction of the photons emitted by the sample, because while photons are emitted in all directions, the detector will occupy only a certain

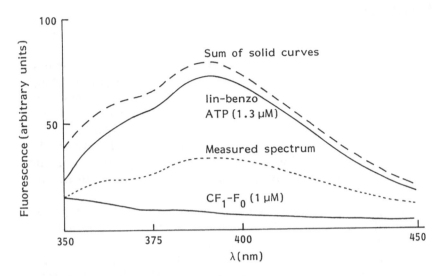

FIGURE 7.1: *Emission spectrum of lin-benzo ATP (a fluorescent analog of ATP) with chloroplast ATP synthase (CF$_1$-F$_0$). λ_{ex} = 332 nm. Note the light scattering due to CF$_1$-F$_0$ (increasing towards lower wavelength) distorts the spectrum of lin-benzo ATP. This is particularly the case as binding quenches the fluorescence of lin-benzo ATP (- - - as compared to ⋯) and hence the contribution of scattered light becomes greater. Data courtesy of S. Benz.*

position in space. Thus the signal from the detector will be instrument-dependent (depending on the geometry of the system). Fluorescent intensities are therefore typically presented in 'arbitrary units', useful for comparative purposes but not directly transferable between different instruments.

It is possible to determine a measure of fluorescence intensity that is independent of the instrument. This is the fluorescence efficiency or quantum yield Q (see Chapter 1), defined as

total number of photons absorbed/total number of photons emitted.

To calculate this, we need to determine the total number of photons emitted from the sample, rather than just the fraction recorded by the detector. This is most simply done by using a standard fluorophore (often provided as a doped plastic block by the manufacturer of the instrument), with known quantum yield, which allows a correction factor, Z, to be measured. However, for most purposes in biochemistry (which involve comparative measurements on a single instrument) this correction is unnecessary.

7.9 Absolute spectra

As outlined in Section 3.2, the lamps used to illuminate a sample in a spectrophotometer or fluorometer will not emit light of equal intensity at all wavelengths. This is particularly true of the intense xenon arc lamp used in some fluorometers. Furthermore, the sensitivity of a photomultiplier (detector) will also vary with wavelength. Thus the current actually recorded by a detector over a range of wavelengths will depend not only on the properties of a sample, but also on the characteristics of the lamp and detector over this wavelength range.

When absorption spectra are measured in modern instruments, these instrumental factors are eliminated by automatic variations in slitwidth and amplifier gain, commonly governed by parallel measurements taken through a reference ('blank') solution (Section 4.3). Such corrections are rarely automatic in commercial fluorometers, where slitwidths are user defined and single beam operation is the norm. Thus fluorescence emission (and excitation) spectra taken on conventional fluorometers are typically distorted (the peaks being skewed in one or other direction) in a manner dependent on the light source and detector characteristics.

Again, this is of little consequence for most (comparative) work performed on a single instrument. However, for some purposes, an instrument-independent 'absolute' spectrum is required. This can be calculated by comparing the measured spectrum, at each wavelength, with the spectrum of a standard whose absolute spectrum is known, and applying the appropriate correction factors. Alternatively, some fluorometers (e.g. Perkin-Elmer LS 5 Series) can operate in a split-beam mode, where a small fraction of the illuminating light is passed directly to a detector (see *Figure 4.9*). The signal from this detector at any wavelength provides the appropriate correction factor, which is then automatically applied to the sample measurement. The effect of such correction is shown in *Figure 7.2*.

7.10 The effect of temperature

Unlike absorption, fluorescence emission is strongly temperature dependent, even between 5°C and 40°C. This is because the processes which quench fluorescence normally involve intermolecular collisions, which increase in rate as the temperature rises. Thus fluorescence

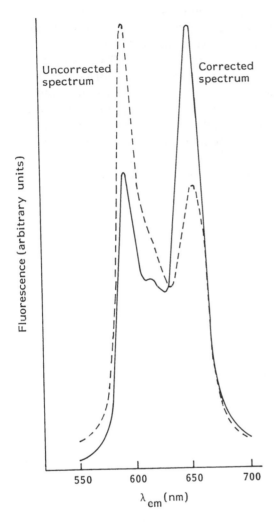

FIGURE 7.2: *Fluorescence of coproporphyrin in HCl (- - -) measured in an instrument with lower sensitivity (lamp emission/detector response) towards higher wavelengths); (—) corrected for instrumental response. Note the λ_{max} values are unchanged but the measured intensities are distorted. Data from Howerton (1967)* Fluorescence – Theory, Instrumentation and Practice *(G. Guilbault, ed.), pp. 233–254. Edward Arnold, London.*

efficiency tends to fall with increasing temperature. Consequently, control of temperature is critical during fluorescence measurements and it is always advisable to use pre-warmed solutions and thermostatted cuvette holders for fluorescence measurements, rather than to rely on a constant room temperature.

As is the case with absorbance measurements, fluorescence can also be measured in frozen solutions using the type of cell holder described in Section 6.4. However, as frozen solutions are not always optically clear, and molar absorptivity rises sharply below −40°C, leading to increased inner filter effects, such fluorescence measurements are normally made using front face detection (Section 4.9). An example of a low temperature fluorescence spectrum is shown in *Figure 7.3*. Comparing the room temperature and low temperature spectra of chloroplasts, we observe that at low temperature fluorescence is seen from the chlorophyll molecules of photosystem I, while at room temperature, this fluorescence is almost entirely quenched.

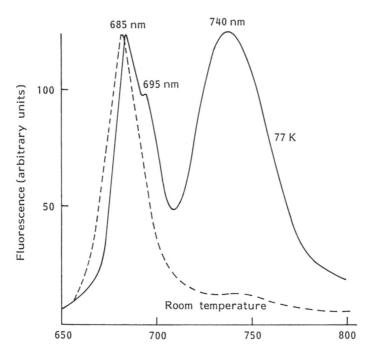

FIGURE 7.3: *Effect of temperature on chloroplast fluorescence. The emission spectra were measured on pea chloroplasts (λ_{ex} = 440 nm) suspended at 20° C (- - -) or 77 K (—). Note that at low temperature (i) the peak at 685 nm shifts towards the red and (ii) the peak at 680 nm is resolved to show two peaks corresponding to chlorophylls of photosystem II, and (iii) the peak at 740 nm (due to photosystem I chlorophylls) is greatly enhanced due to decreased quenching. Reproduced from Hipkins and Baker (1986) in* Photosynthesis: a Practical Approach, *p. 83, by permission of Oxford University Press.*

8 Measurement at a Fixed Wavelength

8.1 Introduction

Determination of an absorption, or fluorescence, spectrum allows us to identify particular chromophores in biochemical systems. Another common use of the spectrophotometer or spectrofluorometer in the biochemistry laboratory is to quantitate (measure the quantity of) various compounds. The procedure used for quantitation is known as a spectrophotometric (or fluorometric) assay.

We can distinguish two types of assay – a static assay, where measurement is made at a single time point, and a dynamic assay, where continuous measurements are made over a period of time. Usually, a static assay will measure the amount of some material present (DNA, protein, ATP etc.) while a dynamic assay will measure a rate of conversion of one compound into another (as, for example, when investigating the activity of an enzyme). Neither routinely requires the measurement of a whole spectrum; measurement is normally carried out at a single (or perhaps two) convenient wavelengths. This can be seen by considering the nature of the absorption, or fluorescence, process itself.

Quantitation by absorbance measurements is based on the Beer–Lambert law (see Chapter 1, Section 1.5), which states that

$$A_\lambda = \log (I_0/I) = \varepsilon_\lambda \, c \, l$$

where ε_λ is the molar absorptivity of the compound under study at the wavelength λ. It can be seen, therefore, that if the absorbance A is measured at a single wavelength, λ, the concentration of the material, c, can be calculated if ε_λ (and the path length of the cuvette, l) is known.

For fluorescence measurements, the analogous relationship is

$$F_\lambda = K c$$

where F_λ is the fluorescence intensity at a single wavelength, λ, and c is the concentration to be determined. K, in this equation, is a composite constant which includes instrumental factors (geometry, lamp intensity, etc.), and the fluorescence efficiency of the fluorophore under study (see Chapter 1). Thus, K for a particular fluorophore is a constant for a given instrument, under constant measuring conditions and, once determined empirically, can be used to calculated fluorophore concentrations under these conditions. Determination of K, for the instrument and the conditions used, is known as the calibration of the assay. Note that K, unlike ε_λ, varies from instrument to instrument, as it depends on the geometry of the cuvette and detecting photomultiplier (Section 7.8).

As discussed in Section 7.3, this linear relationship between F_λ and c may break down at high concentrations, due to the inner filter effect and/or self association of the fluorophore molecules. Care must be taken that assays are performed at concentrations of fluorophore where this relationship holds.

The features required from any assay are given in *Table 8.1*. Perhaps the most striking feature of a biochemical assay is its specificity. As in

Table 8.1: *Requirements of an assay system*

Accuracy	Should give a value close to the true value of the parameter under study	Affected by systematic errors, e.g. use of incorrect value for ε_λ, use of inappropriate (too high) concentrations
Precision	Should be reproducible, with a small standard error in repeated readings	Affected by random errors, e.g. pipetting volumes too small, high background to be subtracted from signal
Sensitivity	Should give measurable values (signal value much higher than machine noise) with amounts of material available	Affected by values of ε_λ, or F_λ, light intensity (slitwidth) and machine response
Specificity	Should give observable signal only when material under study is present. No other compounds should produce a signal or interfere with color development	Affected by wavelength chosen for study, systematic variations in background signal (especially due to light scattering), specificity of enzymes used to develop color
Convenience	Should be rapid and cheap	Use of enzymes increases cost of assay – use of more enzyme costs more but speeds up the process. Increasing the temperature or a change in pH may increase reaction rate

any spectrophotometric assay, specificity is achieved by choosing a wavelength that is characteristic of the compound under study. However, it should also be noted that specificity in a biochemical assay can be enhanced by exploiting the high selectivity of enzymes. These catalysts will act specifically on only one out of a mixture of very similar compounds, and if this reaction generates a chromophore (for example, the generation of a green dye by glucose oxidase in the presence of glucose), a spectrophotometric assay will gain an extra dimension of specificity. Thus enzyme-based assays can be used to quantify one compound in a complex mixture of similar compounds without the need to separate the components of the mixture (*Figure 8.1*).

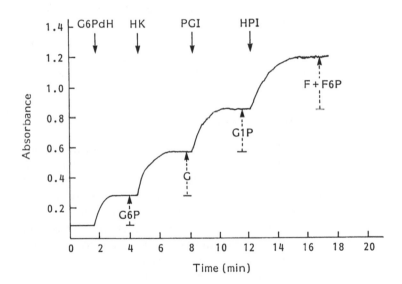

FIGURE 8.1: *Sequential assay of hexose metabolites. The reaction mixture contained about 30 nmol each of glucose (G), glucose 1-phosphate (G1P), glucose 6-phosphate (G6P), fructose (F) and fructose 1-phosphate (F1P), plus the cofactors ATP (4 μmol) and NADP+ (1 μmol). Glucose 6-phosphate dehydrogenase (G6PdH) will generate NADPH (absorbs at 340 nm) only from G6P (first addition). The additional enzymes generate G6P from, specifically, G (hexokinase, HK), G1P (phosphoglucoisomerase, PGI) and F + F1P (hexose phosphate isomerase, HPI), and hence, in the presence of G6PdH, generate further NADPH. Reproduced from Harris and Bashford (eds) (1987) Spectrophotometry and Spectrofluorimetry: a Practical Approach, p. 72, by permission of Oxford University Press.*

8.2 Assay design

8.2.1 Clarifying the solution

Material extracted from tissues, blood, etc., is often inhomogenous and forms cloudy suspensions. For absorbance and fluorescence measurements, these extracts should be clarified, to produce optically clear solutions. Clarification is often combined with a limited purification of the extract to remove interfering materials. How this is done depends on the substance to be measured. For example, when assaying small molecules, protein and nucleic acids can be removed from an extract by precipitating them with acid, followed by filtration or centrifugation. Conversely, small molecules can be removed from an extract by dialysis, protein can be removed from DNA by phenol/chloroform or solid phase extractions, and so on. If assays must be carried in turbid suspensions (e.g. if we wish to quantitate cytochromes within functioning membranes) then difference spectroscopy (preferably dual wavelength spectroscopy) must be used (Section 4.8).

8.2.2 Selecting the wavelength

The wavelength of choice for an assay is that corresponding to an absorption/emission peak, λ_{max}. The peak need not be the highest in the spectrum; it is more important that the peak should be well resolved from wavelengths where interfering materials may absorb light, fluoresce, or scatter light. Thus wherever possible, the peak for measurement is chosen to be in the visible or near ultraviolet region (above 250 nm) where other components of the system (buffers, cuvette material, other extracted compounds) are less likely to absorb.

A peak wavelength is chosen for two reasons. First, the sensitivity of measurement is highest at a peak; here, ε_λ and thus a lower concentration, c, is required to obtain a measurable absorbance of fluorescence. Less obviously, the accuracy of measurement is also maximized. This can be seen from *Figure 8.2*. Close to a peak, a small error in selecting the measuring wavelength ($\delta\lambda$) will lead to a smaller change in absorbance (δA) than it would elsewhere in the spectrum. [This follows from the fact that, at a peak, $dA/d\lambda = 0$ (see Section 6.8)]. To put this another way, close to a peak, the molar absorptivity corresponding to the wavelength actually used $\varepsilon_{(\lambda + \delta\lambda)}$ is approximately equal to the value (ε_λ) used in calculating the concentration, whereas it may be significantly different, say, on the side of an absorption peak where the slope of the curve is greater (*Figure 8.2*).

This rule may be relaxed if the absorbance of the solution is too high for accurate measurement. Choice of an alternative wavelength,

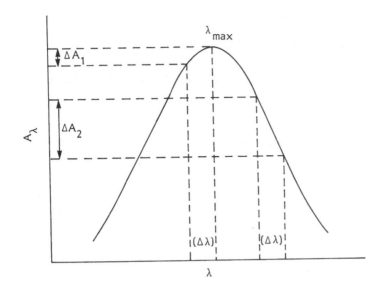

FIGURE 8.2:   *Selection of absorbance wavelength for measurement. A constant error in wavelength (Δλ) leads to a large error in absorbance (ΔA₂) when measurement is on the side of an absorption band, but only a minor change if measurement is at the peak of absorbance (ΔA₁). Reproduced from Harris and Bashford (eds) (1987)* Spectrophotometry and Spectrofluorimetry: a Practical Approach, *p. 72, by permission of Oxford University Press.*

where ε_λ is lower, may allow measurement to be made over a more convenient range. Note, however, that use of an off-peak wavelength will compromise accuracy. For absorbance, therefore, measurements are almost invariably taken at a peak wavelength with the sample being diluted, or placed in a cuvette of shorter pathlength, if necessary.

For fluorescence measurements, however, it may be necessary to excite at an off-peak wavelength, either to avoid the inner filter effect (Section 4.9) or to minimize interference from light scattering (see Chapter 1, Section 1.8). Fluorescence is then detected at an emission peak for the reasons given above. The loss in accuracy that may be experienced here is generally counterbalanced by (i) the increased sensitivity of fluorescence measurements in general, and (ii) the requirement to calibrate fluorescence assays individually (because of their sensitivity to instrument geometry, temperature, etc.) rather than to rely on a literature value for F_λ.

If λ_{max} values for the material under study are not known, then they should initially be determined by measuring an absorption and/or emission spectrum as described in Chapters 6 and 7.

8.2.3 Selecting the concentration

Assays should be performed in regions of sample concentration where (i) the absorption or fluorescence of the solution is proportional to its concentration, and (ii) where sufficient light reaches the detector to keep the signal/noise ratio acceptably high.

In the case of absorption measurements on a standard laboratory instrument, this means diluting the sample/adjusting the light path so that the absorbance lies, ideally, in the range 0.1–1.0. Above $A = 1$, less than 10% of the incident light reaches the photomultiplier and noise levels may rise (and/or stray light becomes a more significant problem – see Section 4.1). Below $A = 0.1$, the difference in intensities between incident and transmitted beams becomes relatively small and hence difficult to measure precisely. Note, however, that this range can be extended at either end using more expensive (double beam or dual wavelength) instruments which rely on a balance between two beams to set a zero value.

In the case of fluorescence measurements, the concentration of sample must be kept low enough to avoid inner filter effects, while being high enough to provide an adequate supply of photons to the detector. In principle, fluorescence measurements are inherently more sensitive than absorption measurements. In absorption at the low concentration limit, we are detecting a small attenuation in a high light intensity (a small percent decrease in photon flux) while in fluorescence, at the low concentration limit, we are detecting a small number of photons arriving at the detector over zero in the absence of sample (an infinitely high percent increase in photon flux). Thus the concentrations used in fluorescence measurements even using standard laboratory instruments are typically 2–3 orders of magnitude lower than those used in absorption measurements – and they can be lowered further if a photon counting mode (digital, rather than analog) fluorometer is available (see Section 2.5).

For a commonly used reagent, we can consider the assay of NADH. The absorption spectrum of NADH is shown in *Figure 8.3a*, with a convenient $\lambda_{max} = 340$ nm, and $\varepsilon_{340} = 6.2 \times 10^3$ M^{-1} cm^{-1}. (Absorption in the far ultraviolet is generally ignored in assay design, as many compounds absorb in this region – see Chapter 1.) Using the above

---

FIGURE 8.3: (a) Absorption spectra of oxidized and reduced NAD⁺ (56 μM), showing large difference in absorbance at 340 nm ($\varepsilon_{340} = 6.22 \times 10^3$ M^{-1} cm^{-1} for NADH). (b) Emission spectrum of 2 μM NADH in 0.1 M phosphate buffer pH 7.0. $\lambda_{ex} = 340$ nm.

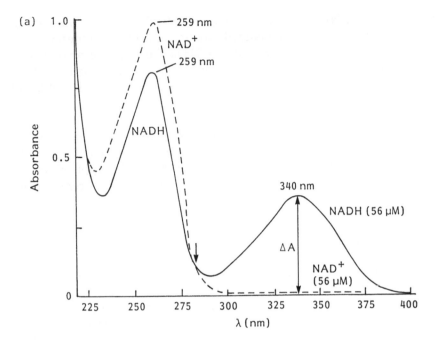

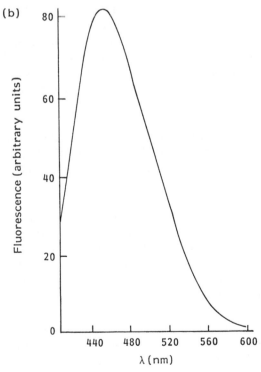

rule of thumb, NADH concentrations of 10–200 µM ($A = 0.06 - 1.2$) can be conveniently measured using spectrophotometry. The fluorescence emission spectrum of NADH is shown in *Figure 8.3b*. λ_{max} = 460 nm. From the height of the peak observed, it is clear that we can readily determine NADH concentrations between 0.1 and 2 µM using fluorometry.

Table 1.1 shows the absorptivities of various compounds of biochemical interest. Typically, absorption measurements may be adequate to determine concentrations down to approximately 1–10 µM; below these concentrations, fluorometric measurements are generally used.

8.2.4 The reference cell

It is essential to refer absorption or fluorescence measurements to a 'reference' or 'blank' cell containing none of the material under study. For single beam measurements, this is read prior to the sample, and any reading obtained subtracted from the sample. For double beam measurements, the reference and sample cells are read simultaneously, and only the difference recorded. In the case of absorbance measurements, the signal obtained with the reference cell defines I_o, the intensity of the incident beam available to the sample material. In modern instruments, the reference cell is placed in the light beam and the resultant detector current is defined as zero absorbance (using a 'zero adjustment' knob or button) (see Chapter 5, Section 5.1).

The reference cell should contain all materials excluding the sample to be measured. Thus, the cuvette material should be identical to that containing the sample, its pathlength should be the same, and any buffer or other solution components (excluding the sample under study) should be identical. For a solution where the color develops (or fades) with time (see below), the reference cell should be prepared at the same time as the sample and incubated under identical conditions. In other words, the reference cell should be prepared and treated in exactly the same way as the cell containing the sample. It is not sufficient to use a plastic cuvette filled with water as a reference for whatever sample is being investigated.

The importance of using a suitable reference cell cannot be too strongly emphasized – it is frequently the major source of error in spectrophotometric assays. Since it is, effectively, subtracted from all samples, an error in the reference cell will lead to a systematic error (i.e. one that cannot be detected by statistical analysis) in all measurements taken. It is always advisable to prepare a reference cell in duplicate or triplicate, and compare absorbances so that an error in its preparation can be readily detected.

If the reference cell itself shows a very high absorption or fluorescence relative to a cuvette containing water, it may be necessary to re-design the assay. Common problems of this type arise (i) if the cuvette material is inappropriate for the wavelength used (see Chapter 3, Section 3.4); (ii) if the solutions used are turbid (see Chapter 1, Section 1.8); or (iii) if the buffer contains some material other than the sample which is strongly absorbent/fluorescent. These problems are often solved easily by using a quartz rather than glass cuvette, or by changing the composition (or batch) of the buffer solution. If, however, the high value of the reference is intrinsic to the system under study (e.g. due to light scattering in a cell suspension), then a variation in technique (to dual wavelength spectrophotometry or front face fluorometry) should be considered. In all these cases, preliminary measurements on a reference cell may identify and eliminate problems without wasting valuable sample.

It is possible that a sample (for example a tissue extract) may contain a number of components in unknown amounts, which cannot be precisely duplicated in a reference cell. Examples include the presence of phenol in DNA extracted from cells using phenol/chloroform, and the presence of other nucleotides (GTP, ADP, etc.) in extracts to be assayed for ATP. Such interfering materials can be detected by measurements at two wavelengths rather than one (see Chapter 6, Section 6.6), and their effects on quantitation avoided by using a color development assay (Section 8.1) specific for the material in question. Alternatively, but with greater difficulty, further purification can be carried out to remove the interfering materials.

8.2.5 Example 1: Measurement of ATP concentration in solution

Performing the measurements.

(i)   The published spectrum of ATP is given in *Figure 8.4*. This establishes

λ_{max} = 260 nm; ε_{260} = 15.7 × 10³ M⁻¹ cm⁻¹; ε_{280} = 2.7 × 10³ M⁻¹ cm⁻¹

This suggests that measurements should initially be made at 260 nm. A silica or polyacrylate cuvette is selected.

(ii) Two reference cells are set up containing water. The cell faces are cleaned with tissue, and placed in a spectrophotometer set up to measure A_{260} at room temperature. The spectrophotometer is zeroed using one reference cell.

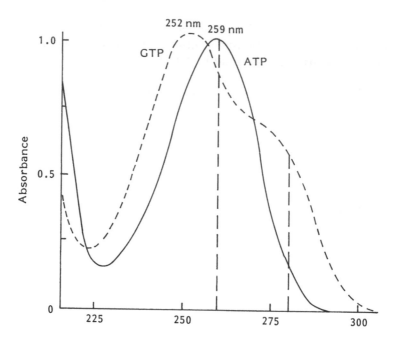

FIGURE 8.4: *Absorption spectra of ATP (65 μM) and GTP (75 μm) at pH 7.0. Note difference in spectral shape and hence in $\frac{A_{280}}{A_{260}}$ ratio. $\frac{A_{280}}{A_{260}}$ = 0.16 for ATP, 0.66 for GTP.*

(iii) The second reference cell is measured. If the deviation is small (< 0.05 absorbance units), it is noted. If it is larger, contamination of one of the cuvettes is suspected and both cells are washed thoroughly. If more cuvettes are to be used in the assay, each is filled with water and the deviation of their absorbances from zero is noted for a 'cuvette correction'.

(iv) A series of dilutions of the sample are made in water until a dilution is found with its absorbance in a measurable range (0.1< A <1). Dilutions of 1/3, 1/10, 1/30, 1/100, 1/300 ..... are convenient to cover a wide range with relatively few samples.

(v) Three replicates of this dilution are made, and three replicates of half (or twice) this dilution, and their A_{260} values are measured precisely. Corrections are made for the deviations measured in (3).

(**NB.** Using two different dilutions of the sample under study allows the operator to check that measurement takes place in the linear range of the assay.)

(vi) The spectrophotometer is set up to read A_{280}, and items (ii) to (v) repeated.

(**NB.** Using a second wavelength for additional measurements allows the operator to check that only compounds with the absorption spectrum of ATP are absorbing light in this region.)

Calculating the result.

(i)   The ratio A_{260}/A_{280} is calculated for the samples. From (i) above, it is known that pure ATP should have $A_{260}/A_{280} = 0.16$. Significant deviations from this value suggest that a second component, with a spectrum different from ATP, is present. If the measured A_{260}/A_{280} ratio does differ appreciably from the theoretical value, a color development assay may be preferable.

(ii) If the measure value of A_{260}/A_{280} is acceptable (0.16 ± 0.01), the concentration (c) of ATP in the original solution can be calculated from

$$c = (A_{260} / \varepsilon_{260}) \times \text{dilution} \quad (\text{mol } l^{-1})$$

(iii) Six values for ATP concentration will be calculated. The estimated value for the original solution will be the mean of these values, $\bar{x}$, with 95% confidence limits given by $(\sigma/\sqrt{n})t$ where n is the number of values determined, and t is the appropriate Student t value.

Note that, for adequate confidence limits, 5–6 readings are required; for lower numbers, $\sqrt{n}$ falls and t rises. Thus if the percent error $(\sigma/\bar{x})$ on a measurement is, say, 5%, 95% confidence limits are $\bar{x} \pm$ 65% if only two readings are taken, but $\bar{x} \pm 12\%$ for six readings.

8.3 Quantitation of nonabsorbing material

8.3.1 Color development assays

As noted in Chapter 2, compounds that are colored, or fluorescent, are rare in biochemistry. Furthermore, compounds that do absorb light in the near ultraviolet, like protein and DNA, have a relatively low absorptivity, making their quantitation by absorbance measurements insensitive. Even when the material under study is chromophoric, other compounds may absorb light in the same region (many compounds with aromatic rings absorb in the near ultraviolet, for example) making absorbance measurements nonspecific.

These problems may be overcome by a color development assay. Here the compound under test reacts specifically with a reagent to form a colored product which can then be measured. The reaction may be noncovalent – as when a macromolecule interacts with a dye – or it may involve chemical changes in measured compound, causing it to generate a chromophore. In the latter case, specificity may be achieved by using an enzyme to catalyze the chemical change.

8.3.2 Dye binding assays

Dye binding assays are now widely used for the quantitation of both protein and nucleic acids. For proteins, a convenient dye is Coomassie brilliant blue® G250. This is pale orange in acid solution (protonated form), but changes to bright blue when bound to protein. The dye (in phosphoric acid solution) is simply added to a protein solution, and the color measured at 595 nm. As an indication of sensitivity, $A_{595} = 1$ in this assay is equivalent to about 20 µg ml^{-1} protein; this can be compared with $A_{280} = 0.7$ for 1 mg ml^{-1}, when near ultraviolet absorbance is measured directly on the protein solution.

For DNA, a convenient dye is Hoechst 33258 (*bis*-benzimide H33258). This produces a highly fluorescent product ($\lambda_{ex} = 355$ nm, $\lambda_{em} = 460$ nm) when added to a solution of DNA, which can thus be quantitated in the 0.1–1 µg ml^{-1} range. Again, this can be compared with direct ultraviolet measurements, this time at 260 nm, which are applicable in the 5–50 µg ml^{-1} range. Furthermore, the dye produces virtually no fluorescence with RNA, and thus the fluorescence assay is specific for DNA alone.

8.3.3 Enzyme-based assays for metabolites

In an enzyme-based assay, a metabolite is reacted with some compound to produce a fluorophore or chromophore, the reaction being catalyzed by a specific enzyme. Alternatively, a chromophore may be supplied which is used up during the reaction.

A large number of metabolite assays (see *Table 8.2*) involve dehydrogenases, which oxidize the metabolite using the cofactors NAD$^+$ or NADP$^+$ and thus produce the reduced cofactor NAD(P)H which absorbs light with $\lambda_{max} = 340$ nm, and fluoresces at about 460 nm. (*Figure 8.3*) (In some cases, NAD(P)H may be oxidized in the assay; the direction observed is governed by ΔG for the reaction.) In all cases the color change observed is the same because the component observed is the cofactor, which is common to all reactions; the enzyme selects the metabolite from the mixture and thus defines the specificity of the assay. Another large group of metabolite assays (*Table 8.3*) use oxidases, which oxidize their specific substrate using O_2 from the air, producing hydrogen peroxide, which in turn is used by a second enzyme, peroxidase, to oxidize a dye precursor (leuko-dye) to a colored compound (see Chapter 2, *Figure 2.10*).

In principle, the concentration of metabolite under study can be calculated either from the rate of reaction observed or, by waiting until all the metabolite is converted, from the overall extent of the reaction (*Figure 8.5*). Rate assays are rarely used, partly because their sensitivity is inherently lower (since measurements are made when only a fraction of substrate has been converted viz. only a fraction of the potential absorbance/fluorescence change has occurred), partly because rates are more difficult to measure precisely than static values, and partly because enzyme catalyzed rates are hyperbolically,

TABLE 8.2:   *Metabolite assays using dehydrogenases*

Analyte	Enzymes used	Reaction sequence
Isocitrate	Isocitrate dehydrogenase	Isocitrate + $NADP^+ \rightarrow$ 2-oxoglutarate + CO_2 + NADPH
Alanine	Alanine dehydrogenase	Alanine + $NAD^+ \rightarrow$ pyruvate + NH_4^+ + NADH
Lactate	Lactate dehydrogenase	Pyruvate + NADH $\leftarrow$ lactate + $NAD^.$ (equilibrium displaced to left with hydrazine)
Glycerate 3-phosphate	Phosphoglycerate kinase Glyceraldehyde 3-phosphate (GAP) dehydrogenase	Glycerate 3-phosphate + ATP $\rightarrow$ glycerate 1,3-diphosphate +ADP Glycerate 1,3-diphosphate + NADH $\rightarrow$ GAP + NAD^+
Coenzyme A	β-Hydroxybutyryl CoA dehydrogenase	CoA + diketene $\rightarrow$ acetoacetyl CoA (non-enzymatic) Acetoacetyl CoA + NADH $\rightarrow$ β-hydroxybutyryl CoA + NAD^+
4-Amino butyrate	4-Aminobutyrate/glutamate transaminase Succinate semialdehyde dehydrogenase	4-Aminobutyrate + 2-oxoglutarate $\rightarrow$ succinate semialdehyde + glutamate Succinate semialdehyde + $NADP^+$ $\rightarrow$ succinate + NADPH
Erythrose 4-phosphate	Transaldolase	Erythrose 4-phosphate + fructose 6-phosphate $\rightarrow$ sedoheptulose 7-phosphate + GAP
	Triose phosphate isomerase	GAP $\rightarrow$ dihydroxyacetone phosphate
	Glycerol 3-phosphate dehydrogenase	Dihydroxyacetone phosphate + NADH $\rightarrow$ glycerol 3-phosphate + NAD^+
UDP	Nucleoside diphosphate kinase	UDP + ATP $\rightarrow$ UTP + ADP
	UDPG pyrophosphorylase	UTP + glucose 1-phosphate $\rightarrow$ UDP-glucose + PP_i
	UDPG dehydrogenase	UDP-glucose + $NAD^+ \rightarrow$ UDP-glucuronate + NADH

Examples of various assays are given; in all cases the production of NAD(P)H is stoichiometrically linked to the amount of compound being assayed. The list is not meant to be exhaustive. Additional examples are given in the text (see *Figure 8.1*).

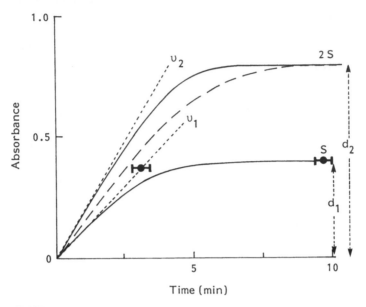

FIGURE 8.5: *Time course of an enzyme-mediated assay, the enzyme reaction generating a colored product (see Figure 8.1). At the end point, twice the amount of reactant has generated twice the color ($d_2 = 2d_1$). However, if the initial rate is measured, the rate is not necessarily proportional to [reactant] ($v_2 < 2v_1$). Also note that an error in time of measurement or amount of enzyme added (− − −) will affect the rate measured but not the end point.*

rather then linearly related to substrate concentrations. End point assays are therefore the most widely used for measuring the concentrations of small organic molecules in biochemistry.

8.3.4 Example 2: Measurement of ATP concentration in solution

Making the measurement.

(i) The molar absorptivity at 340 nm, ε_{340}, for NADH is 6.2×10^3 M^{-1} cm^{-1}. NAD⁺ has negligible absorption at this wavelength. Measurements are thus made at 340 nm using a cuvette made from polyacrylate or special optical glass. [The wavelength cutoff of normal glass is approximately 330 nm (see Chapter 3, Section 3.4), and thus it has significant absorption at 340 nm.]

(ii) A reference cell is set up containing buffer (to ensure a suitable pH for enzyme activity), $MgCl_2$ (a cofactor for hexokinase),

TABLE 8.3: *Metabolite assays using oxidases*

Analyte	Enzymes used	Reactions generating hydrogen peroxide
Glucose	Glucose oxidase	Glucose + O_2 + H_2O → gluconolactone + H_2O_2
Uric acid	Uricase	Uric acid + O_2 + $2H_2O$ → allantoin + H_2O_2
Cholesterol	Cholesterol oxidase	Cholesterol + O_2 → Δ^4-cholestenone + H_2O_2
2-Oxoglutarate	Glutamate–pyruvate transaminase	2-Oxoglutarate + alanine → pyruvate + glutamate
	Pyruvate oxidase	Pyruvate + PO_4^{3-} + O_2 → acetyl phosphate + H_2O_2
Triglycerides	Esterase	Triglycerides + $3H_2O$ → glycerol + 3 fatty acids
	Glycerokinase	Glycerol + ATP → glycerol 3-phosphate + ADP
	Glycerophosphate oxidase	Glycerol 3-phosphate + O_2 → dihydroxyacetone phosphate + H_2O_2
Creatinine	Creatininase	Creatinine + H_2O → creatine
	Creatinase	Creatine + H_2O → sarcosine + urea
	Sarcosine oxidase	Sarcosine + O_2 + H_2O → glycine + HCHO + H_2O_2

Examples of various assays are given; in all cases the production of H_2O_2 is stoichiometrically linked to the amount of compound being assayed, and this is used, by the enzyme peroxidase, to oxidize a leucodye (e.g. 2,2′-azino-di-[3′-ethylbenz thiazoline sulphonate] (ABTS®) or tetramethylbenzidine) to a colored compound (see Section 2.6).

glucose and NAD⁺ (substrates) and glucose 6-phosphate dehydrogenase (NAD⁺–linked, from *Leuconostoc*). This is used to zero the spectrophotometer. Measurements are made at 37°C (to speed up the enzyme reaction). Glucose and NAD⁺ concentrations should be higher than (in excess of) the highest concentration of ATP likely to be used, so that only the amount of ATP will limit the extent of reaction.

(iii) A sample cell is set up, containing in addition, an aliquot of the unknown solution of ATP. [A series of trial experiments, with a series of ATP dilutions, may be necessary to obtain a final absorbance between 0.05 and 1 (see example 1).] Its absorbance is recorded. Deviation from the reference value should be small (< 0.05 units) unless the unknown solution contains other absorbing compounds.

(iv) Hexokinase is added to both cuvettes, and the absorbance of each followed in time (see *Figure 8.1*). A_{340} for the sample cell should rise as NADH is produced according to the reactions

hexokinase

glucose + ATP → glucose 6-phosphate

G-6-P dehydrogenase

glucose 6-phosphate + NAD⁺ → 6-phosphogluconate + NADH

When the absorbance shows no further change in time, it is recorded.

(v) Replicates (six, at two different dilutions of ATP) are performed as in Example 1.

(vi) The difference ($\Delta A_{sample} - \Delta A_{reference}$) corresponds to the absorbance change due to the presence of ATP. Since one mol NADH is produced per mol ATP used, the original concentration of ATP can be calculated from

$$c = \frac{(\Delta A_{sample} - \Delta A_{reference})}{6.2 \times 10^3} \times \frac{\text{vol of reagents}}{\text{vol of ATP added}} \times \text{original ATP dilution mol l}^{-1}$$

Notes on this assay. We can compare this procedure with that described in Example 1.

(i) The sensitivity of the assay in Example 2 is some 2.6-fold less if absorbance measurements are taken ($\varepsilon_{260}^{ATP} = 16.7 \times 10^3 \text{ M}^{-1}\text{cm}^{-1}$, while $\varepsilon_{340}^{NADH} = 6.2 \times 10^3 \text{ M}^{-1} \text{ cm}^{-1}$). However, since NADH is fluorescent, the sensitivity of the assay in Example 2 can be much higher if fluorescence measurements are taken.

(ii) The specificity of the assay in Example 2 is higher. Hexokinase will utilize only ATP, so any irrelevant absorbing molecules will be unchanged during the assay. Since only an absorbance change is measured, such molecules will not affect the reading obtained. This is true not only of buffer components – which would at least be detected in Example 1 by measuring at two wavelengths – but also of compounds such as ADP, and AMP, which have identical absorption spectra to ATP and would (undetectably) raise the apparent concentration of ATP, in Example 1, above its true value.

(iii) Since 'before' and 'after' measurements are made in a single cuvette, each cuvette serves as its own reference cell, further decreasing variability.

8.4 Calibration

For defined compounds (ATP, NADH, chlorophyll etc.), molar (or specific) absorptivities (ε_λ) are generally available from the literature. To calculate concentration from an absorbance measurement is therefore a simple arithmetical manipulation (see Examples 1 and 2). The situation is less straightforward for fluorescence assays. In these cases, the observed signal depends not only on the inherent fluorescence of the sample, but also on the geometry of the system (only a fraction of the emitted photons being detected (see Chapter 7, Section 7.8), and, markedly, on environmental conditions [temperature, presence of oxygen, etc. (see Chapter 2, Section 2.4)]. Literature values of quantum yield, therefore, are not readily related to the fluorometer signal recorded in an assay. Instead, the instrument is calibrated empirically; the response of the instrument to amounts of the material under study is determined using a range of (standard) solutions, and the signal due to the unknown sample converted to a concentration using this, hopefully linear, relationship. There are clearly errors involved in reading concentrations from a calibration curve; these are difficult to analyze statistically but, as a convenient rule, accurate determinations can be made only if the reading obtained from the unknown sample lies close to the center of the calibration curve.

For fluorescence measurements (e.g. of NADH), it is often convenient to measure concentration in a 'stock' solution accurately by absorption measurements, to prepare precise dilutions from this known solution, and to use these dilutions to calibrate the fluorometer. Alternatively, the standard concentrations may be determined by weighing. This may seem, *a priori*, the most obvious procedure, but it is not always accurate for biochemicals which are commonly hygroscopic. Variable amounts of water may thus be weighed out along with the compound under study, leading to an overestimate of its true concentration in the calibrating solution.

The calibration procedure is also followed for absorbance measurements where ε_λ is not known, or may be dependent on the conditions employed. Examples include the dye binding assays described in Section 8.3. Again, a calibration curve is constructed using known dilutions of a standard solution and the unknown sample compared with these values.

A summary of the procedure followed in designing an assay is outlined in *Figure 8.6*, and the considerations involved in assessing its effectiveness are given in *Table 8.1*.

8.5 Additional considerations for fluorescence assays

Under ideal conditions, the fluorescence of a solution will increase linearly with the concentration of the fluorophore. We have seen above that this relationship holds true only if environmental conditions (notably temperature and solvent composition) are kept constant and if the absorbance of the solution is sufficiently low that the inner filter effect is negligible.

However, other factors may affect the measured fluorescence. Various molecules (notably O_2, but also I^-, some metal ions, etc.) will quench fluorescence by colliding with the excited state and carrying off energy as heat. Thus these molecules promote nonradiative decay of the excited state and the observed fluorescence is lowered (see Chapter 1, Section 1.6). For consistent fluorescence measurements (i.e. for valid comparison of an unknown with a calibration solution), it is therefore important to remove these quenchers or, in the case of oxygen, at least maintain a constant level (as in air saturated buffer at a fixed temperature). While this condition might seem straightforward, it is frequently breached; mixing an aqueous solution at 25°C with an aqueous solution at 0°C, or a nonaqueous solution at 25°C, will significantly change the concentration of dissolved oxygen.

Another factor that can effect fluorescence measurements is the photochemical destruction of the fluorophore in the light beam (photobleaching). This is observed as a slow decline in signal during continuous illumination, and hence makes reproducible measurements difficult.

Photobleaching can be largely avoided by using an instrument with a pulsed light source (see Chapter 3, Section 3.2), where the total illumination time is only a fraction of the time of observation. If the fluorometer employed has a high intensity, continuous light source, its effects can be minimized by (i) interrupting the illumination, using a shutter, when measurements are not being made, and (ii) stirring the solution. Stirring is effective, perhaps contrary to expectations, as only those molecules in the light beam (a small fraction of the total) are bleached; stirring replaces these bleached molecules with undamaged ones from the bulk of the solution.

8.6 Assays in turbid solutions

Notwithstanding the above caveats, the proportionality between concentration and absorbance/fluorescence holds well for optically clear solutions and can be used readily for quantitation. This is not true for turbid solutions, which show a high apparent absorbance or fluorescence even in the absence of chromophores, as a result of light scattering (see Chapter 1, Section 1.8).

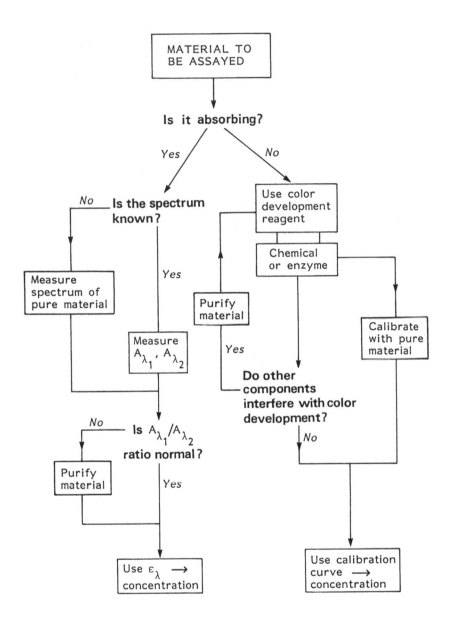

FIGURE 8.6: *Scheme for assay design using spectrophotometry. For fluorometric assays, a similar scheme is followed, but sensitivity can be > 100 × higher (<1 nmol as compared to >50 nmol).*

As described previously, corrections can be made for turbidity by measuring at two wavelengths, one where the chromophore does not absorb (the 'reference' wavelength) and one at (or close to) an absorbance/fluorescence peak (the 'measuring' wavelength). By subtracting the reference signal from the measuring signal, the effects of turbidity can be avoided. In fact, as dual wavelength fluorometers are rare, absorbance measurements are normally used for assays on turbid solutions.

For example, the quantitation of cytochrome c in mitochondria can be accomplished using a measuring wavelength of 550 nm, and a reference wavelength of 540 nm. These wavelengths are sufficiently close that light scattering is roughly equal at both wavelengths. However, the absorbance difference between oxidized and reduced cytochrome c is large at 550 nm, but zero at 540 nm (an isosbestic point for the reduction) (*Figure 6.7a*).

Absorbance is measured on oxidized mitochondria at a known concentration (say 1 mg ml^{-1}) at 540 nm and 550 nm, using a dual wavelength spectrophotometer. While both signals are high (*Table 8.4*), due to light scattering, the difference is very small and can be determined accurately only with a dual wavelength instrument. The cytochrome is then reduced (using ascorbate), and the change in absorbance at each wavelength is measured. Any changes at 540 nm represent changes in light scattering, which will be common to both wavelengths. However, changes in the difference ($A_{550} - A_{540}$) should reflect only changes in the redox state of the cytochrome c.

Since $\Delta\varepsilon_{550}$ (the change in molar absorptivity at 550 nm during reduction) is known, from studies on cytochrome c in free solution, the concentration of cytochrome c in this turbid suspension can be calculated.

TABLE 8.4: *Measurement of cytochrome c concentration in a mitochondrial suspension*

	A_{540}	A_{550}	Signal ($A_{550} - A_{540}$)	ΔA_{550}
Before reduction	1.8251	1.7926	−0.0325*	0.0000
After reduction	1.6827	1.6590	−0.0237	+0.0088
Calculation of concentration	$\Delta\varepsilon_{550} = 2 \times 10^4$ M^{-1} cm^{-1}			$c = 0.44$ µM

* Reference value – defined as zero.

8.7 Spectrophotometric measurement of rates

8.7.1 Problems in determining rates

We have seen above how concentrations in solution can be measured by spectrophotometry or spectrofluorometry, by making individual measurements at one (or two) wavelengths or, in cases where a color is developed from an otherwise nonabsorbing sample, by making measurements before and after the color development.

Such techniques are readily adapted to measuring the rate of conversion of a reactant to a product provided either the product or reactant is colored or fluorescent. A suitable wavelength (typically λ_{max}) is selected and measurements taken 'continuously' rather than at an arbitrary time or times. The considerations described above for static measurements (linear relationship between signal and concentration, problems with turbidity, etc.) are equally applicable to rate measurements.

However, because during rate measurements the signal will be changing over the period of measurement, additional factors must be considered. These include methods of defining a 'time zero' (i.e. how to start the reaction), recording a measurement very rapidly (so that the absorbance/fluorescence change over the time taken to make a measurement is negligible), or alternatively, stopping the change very rapidly before measurement. Furthermore, since reaction rates are very sensitive to temperature, the temperature at which the change is studied must be carefully controlled, by thermostatting, even in the case of absorbance measurements.

8.7.2 Conventional measurements of enzyme rates

In biochemistry, most rate measurements are made on enzyme-catalyzed reactions. Such measurements may be performed to estimate how fast the enzyme might be working in the cell, when tissue extracts would be used as the enzyme source. Alternatively, they may be used to study the substrate, inhibitor or pH dependence of the enzyme, in order to understand better its active site, in which case purified enzyme might be used. In either case, the rate measured should be proportional to the amount of enzyme added, and it is the first duty of the experimentalist to establish this proportionality (and, if necessary, to adjust conditions so that it occurs). [If the measured rate is proportional to the amount of enzyme added, its specific activity (mol substrate converted per min per mg enzyme) will be constant over the measured range.] Examples of deviations from such linearity may occur if the enzyme dissociates at low concentrations, in which case rates are underestimated at low enzyme concentrations or if the rate is too fast to measure accurately, in which case rates at high enzyme concentration are underestimated.

Once linearity of response has been established the operator can decrease the rate to be measured to a convenient level (< 1 absorbance unit per min) by decreasing the amount of enzyme. If such a rate can be achieved within the linear range of the assay, manual mixing methods and pen based data collection methods are adequate; if not, specialized (and hence more expensive) rapid mixing devices and rapid data collection are necessary (see Chapter 9, Section 9.3). The progress of an enzyme assay is summarized in *Figure 8.7*.

8.7.3 Manual mixing devices

Typically, a cuvette containing all necessary reagents bar one is prepared, and equilibrated at the assay temperature. The missing reagent should be the least stable under the assay conditions, and is typically the enzyme solution itself. To start the reaction, a small volume of this reagent is pipetted on to a glass or plastic 'plumper' (*Figure 5.3*). This is then introduced into the cuvette, and simultaneously agitated to stir the solution. Mixing is complete within about 10 sec.

8.7.4 Data collection

During the reaction, recording the absorbance/fluorescence change from a digital display is difficult (although possible) as the display is constantly changing. More conveniently, an analog output from the spectrophotometer is led to an x–t (chart) recorder where the pen displacement (x) – representing the absorbance – is recorded, on a moving paper roll, as a function of time (t). The time resolution of this system depends on the speed of mechanical movement of the pen (see Chapter 6, Section 6.4), but rates of less than 1 absorbance unit per min are readily accessible.

Alternatively, the signal may be stored in digital form on a magnetic disc, as occurs in many computer-driven instruments. This can increase the time resolution to some extent, but is of more general convenience in that the computer can be readily programed to

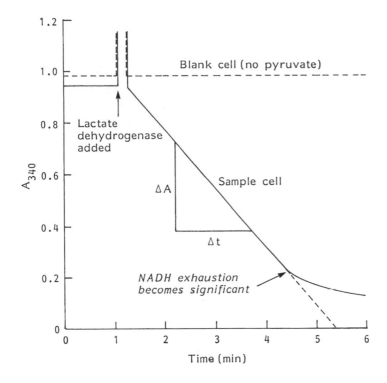

FIGURE 8.7: *Assay of lactate dehydrogenase. The progress curve for NADH oxidation is shown. For details see Section 8.7.5.*

calculate the slope, and thus print out the rate of change, directly. Such programs ('kinetics packages') are commercially available for most current instruments.

8.7.5 Example 1: Assay of lactate dehydrogenase

Making the measurement.

(i)   The reaction catalyzed is

$$\text{pyruvate} + \text{NADH} + \text{H}^+ \rightarrow \text{lactate} + \text{NAD}^+$$

The reaction has a highly negative ΔG, and thus proceeds in the direction shown. NADH absorbs light strongly at 340 nm (ε_{340} = 6.2×10^3 M^{-1} cm^{-1}), while NAD$^+$ does not.

(ii)   Three cuvettes are set up, with contents as in *Table 8.5*. Buffer is included to maintain a pH optimal for enzyme activity. The concentration of NADH is limited to 0.2 mM (absorbance ≈ 1.3) so that the spectrophotometer is operating in a range where signal/noise is low, and the linear relation between absorbance and concentration holds.

(iii) The spectrophotometer is zeroed using Cuvette 1 (no NADH).

(iv) Cuvette 2 (sample) is brought into the measuring beam, and its A_{340} measured over 1–2 min. (This identifies any nonenzymic destruction of NADH – it acts as an 'enzyme blank'.) The rate of change of A_{340} ($\Delta A_{340}/\Delta t$) should be zero or very small; if the rate is perceptible, it must be subtracted from the rate measured in the presence of enzyme to give the enzyme catalyzed rate.

(v) Lactate dehydrogenase (in a volume of <50 µl) is added to Cuvette 2, and ΔA_{340} followed in time (*Figure 8.7*). The change should be linear in time until NADH is nearly completely consumed (A_{340} < 0.1). The rate is taken as ($\Delta A_{340}/\Delta t$) over this linear region. If the observed rate is too fast to measure accurately, this step is repeated using less enzyme.

(vi) Cuvette 3 (no pyruvate) is brought into the measuring beam and ($\Delta A_{340}/\Delta t$) measured both before and after the addition of enzyme. It should be very small in both cases. Any rate measured in this cuvette ('substrate blank') is due to oxidation of NADH by contaminating oxidases in the enzyme mixture, and should be subtracted from the result obtained in Cuvette 2 to obtain the rate due to lactate dehydrogenase.

(vii) Six replicates of Cuvette 2 are prepared, and rates measured with the addition of two different amounts of lactate dehydrogenase. The observed rates (after correction for the blank rates) should be proportional to enzyme concentration, and the rate catalyzed by (activity of) a known amount of enzyme can then be calculated.

TABLE 8.5:   Assay of lactate dehydrogenase

	Cuvette 1 (A=0)	Cuvette 2 (reaction cuvette)	Cuvette 3 (substrate blank)
Phosphate buffer (0.1 M, pH 7.0)	2.8 ml	2.8 ml	2.8 ml
Sodium pyruvate (25 mM)	0.1 ml	0.1 ml	0
NADH (20 mM)	0	0.03 ml	0.03 ml
Water	0.1 ml	0.07 ml	0.17 ml
Samples incubated for 5 min at 30°C			
Lactate dehydrogenase	0	0.01–0.05 U*	0.01–0.05 U
Reaction followed at 340 nm			

For explanation, see text.
*1 unit (U) of enzyme is the amount that will convert 1 µmol substrate per minute.

Calculations.

(i)   The true enzyme rate is the difference between the sample and blank values

$$(\Delta A_{340}/\Delta t)_{sample} - (\Delta A_{340}/\Delta t)_{blank} = (\Delta A'_{340}/\Delta t) \text{ absorbance units per min}$$

(ii)   To convert absorbance units to concentration changes, use ε_{340}.

$$(\Delta c/\Delta t) = (\Delta A'_{340}/\Delta t) \cdot 1/\varepsilon \text{ M min}^{-1}$$

(iii)   An amount of enzyme is defined by the amount (number of mol) of substrate it converts per unit time. This is given by the concentration change in the cuvette multiplied by the cuvette volume (v).

$$\text{No. of mol converted per min} = (\Delta A'_{340}/\Delta t) \cdot v/\varepsilon$$

(where $\varepsilon = M^{-1} cm^{-1}$, t = min, v = liters, and cuvette path length = 1 cm).

(iv)   Confidence limits on the values obtained are calculated as in Section 8.2.

(v)   A similar procedure can be used for fluorometric rate measurements. However, if a rate is measured fluorometrically, the fluorometer must be calibrated (with known amounts of NADH in this case) so that the conversion factor from ΔF to Δc is known for the fluorometer being used (see Section 8.4).

9 Probing the Environment

9.1 Environmentally sensitive chromophores

The absorption and fluorescence properties of molecules differ according to their chemical nature. These properties can be used to identify molecules (see Chapters 6 and 7) and to measure their concentrations (quantitation) or the rates of their interconversions (see Chapter 8).

It has also been pointed out, in Chapter 1, that the absorption or fluorescence properties of a particular molecule may vary with its environment. A simple example would be the change in color of a pH indicator with the pH of its environment, and analogous examples (Ca^{2+} indicators, etc.) which have been described in Chapter 2.

More subtle changes are also seen. The fluorescence of a molecule, for example, is often increased (increase in quantum yield) when it is bound to a protein where solvent access to the fluorophore (and hence solvent-induced quenching) is restricted. The absorption spectrum of a linearly conjugated molecule (such as a carotenoid) will shift when it is subject to an electrostatic field. Other examples are given in Chapter 2, Section 2.3.

The sensitivity of absorbance and fluorescence to environment is a disadvantage in the techniques described in Chapters 6–8. It means that for quantitation or identification of a compound from its absorbance or fluorescence, the conditions chosen (buffer, solvent, temperature) must be carefully controlled and standardized so that constant calibration values can be used. However, we can turn this sensitivity to our advantage in that it allows us to use a particular fluorophore or chromophore as a probe of its environment. In this case, changes in its optical properties signal to the observer information about the region where the probe is situated – properties such as its pH, hydrophobicity, etc.

This has been of great importance in biochemistry. If such probes were unavailable, the only way of obtaining information about complex biological systems would be to break them up, separate the components, and study each individually. In contrast, a probe placed in a particular cell component (e.g. the cytoplasm) or membrane (e.g. the mitochondrial inner membrane) can report on events in its immediate environment, without destroying the organization we should like to study (e.g. a distribution of ions between different cell compartments). And, since measurement of absorbance or fluorescence is itself a nondestructive process, these events can be studied continuously over prolonged time periods. Some examples of the use of environmentally sensitive chromophores and fluorophores are described in this chapter.

9.2 Measuring ligand binding to proteins

The binding of small molecules (ligands) to proteins is an important feature of biochemistry. The proteins may, for example, be carriers (as in oxygen binding to hemoglobin), enzymes (as in the formation of an enzyme–substrate or enzyme–regulator complex) or receptors (as for estrogen or adrenaline). The strength of the binding (its affinity), its specificity, and its rate are all important biochemical parameters as they will affect the behavior, and thus the function, of the protein within the cell.

Measuring ligand binding by classical methods (equilibrium dialysis, ultrafiltration) is slow and requires large amounts of material (and often the use of radioactivity). Such methods rely on the difference in size between protein and ligand to allow the separation, and subsequent measurement, of free and bound ligand. If, however, there is an optical change on ligand binding, association can be followed, rapidly and conveniently, using spectrophotometry or fluorometry.

The optical change may be in either the ligand or the protein. Examples of naturally occurring ligands that show such changes on protein binding are few, although NADH (and NADPH) provides a widely exploited example (Section 9.3). Other chromophoric ligands are typically modified forms of naturally occurring ligands, such as the fluorescent ATP analog, shown in Chapter 2.

Changes in the absorbance or fluorescence of the protein on ligand binding are more widespread. Some proteins, like hemoglobin, contain a chromophore (heme) which is directly involved in ligand binding, and undergoes a marked change in absorbance. More generally, nearly all proteins contain tryptophan, a fluorophoric amino acid, which, if situated near a binding site, will signal binding by changing its fluorescence. Tryptophan is perhaps not the ideal environmental probe. Firstly it does not show a wide separation between emission and excitation peaks (*Figure 9.1*), and thus its signal may be limited by interference from scattered light. Secondly, proteins rarely contain a single tryptophan residue precisely at their active sites – more often there are several at different locations within the protein – and thus the signal may be superimposed on a constant background of unchanging fluorescence from these other residues. None the less, its wide distribution has made tryptophan fluorescence a useful monitor of ligand binding in a variety of systems. One example is given in Section 9.3.

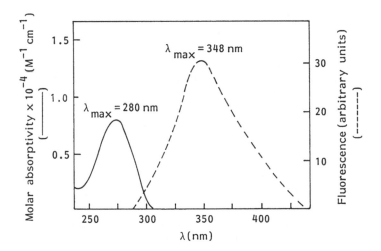

FIGURE 9.1: *Absorption spectrum (—) and fluorescence emission spectrum (---) of tryptophan in water. When present in proteins, the fluorescence emission peak may shift to a lower wavelength (see Section 9.3.3).*

9.2.1 Example 1: Determination of the affinity of lactate dehydrogenase for NADH

Experimental method.

(i)   This method utilizes the change in NADH (ligand) fluorescence when it binds to the enzyme. First it is necessary to establish the absorption and fluorescence properties of NADH (a) when free in solution, and (b) when bound to the protein. λ_{max} for the absorption of free NADH is 340 nm (Chapter 1, Section 1.7). Using this as an excitation wavelength, an emission spectrum of NADH in water is determined as discussed in Chapter 7. This is repeated for NADH in the presence of a roughly equal concentration of protein (see *Figure 2.4*).

(ii)   A difference spectrum is calculated. This shows that, when NADH binds to lactate dehydrogenase:
(a) its fluorescence is enhanced about threefold, and
(b) its emission peak shifts from 460 nm, for free NADH, towards the blue end of the spectrum.

(iii) Conditions for following NADH binding to lactate dehydrogenase can now be selected.
(a) Excitation will be at 360 nm. This lies on the edge of an absorption peak, allowing a wide range of NADH concentrations to be used with a minimal inner filter effect (Chapter 4, Section 4.9).
(b) The emission wavelength will be at 445 nm, where a large difference in fluorescence is observed between bound and free NADH.
(c) A narrow excitation slit (6 nm) is chosen, to maintain selectivity in excitation [see (a)] while the emission slit can be wider (16 nm) to maximize the signal obtained. (As the entire emission peak increases in size during fluorescence enhancement, precise wavelength selection for emission measurements is unnecessary.)

(iv)   A suitable volume of buffer is added to the cuvette, and small aliquots of NADH are added sequentially. After each addition, the mixture is stirred and F_{445} recorded. The solution of NADH should be concentrated so that the entire range of NADH concentrations required can be covered without changing the volume in the cuvette appreciably (< 5%). A plot of F_{445} versus [NADH] (*Figure 9.2,* open symbols) should be linear over the range studied, showing that the inner filter effect is negligible. (If it is not, shift the excitation wavelength even further away from λ_{max} for NADH absorption.)

(v)   A sample containing buffer and enzyme is then titrated with NADH in the same manner. A plot of F_{445} versus [NADH] is initially nonlinear (*Figure 9.2*, closed symbols) as NADH bind to the protein giving a fluorescence higher than that for free NADH.

(vi)   The difference between curves in *Figure 9.2* represents the fluorescence enhancement resulting from NADH binding, and thus represents the binding isotherm for NADH binding to the protein (*Figure 9.3*). The dissociation constant, K_d, for NADH can be calculated from this curve.

Calculation.

(i)   The fluorescence change observed in the difference curve (ΔF_c) is assumed to be proportional to the amount of NADH bound. (If there is a reason for doubting this linear relationship, this assumption should be checked by direct binding studies.) To find the proportionality constant, the value ΔF_∞ (fluorescence change when all binding sites are occupied by NADH) must be determined. When this is known, the fractional saturation, α (α = number of binding sites occupied/total number of binding sites) is given by $\Delta F_c/\Delta F_\infty$.

(ii)   We cannot experimentally increase [NADH] infinitely. However, this can be simulated mathematically by plotting $1/\Delta F_c$ versus $1/$[NADH] and extrapolating back to the vertical axis ($1/$[NADH] $= 0$; [NADH] $= \infty$). This is shown in *Figure 9.3b*.

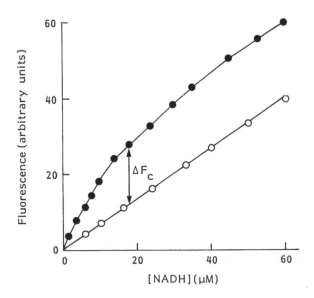

FIGURE 9.2:   *Fluorescence titration of lactate dehydrogenase with NADH.*
$\lambda_{ex} = 360$ nm, $\lambda_{em} = 445$ nm. Open circles represent buffer alone, filled circles show buffer + lactate dehydrogenase (14 µM active sites).

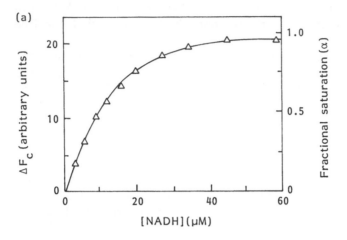

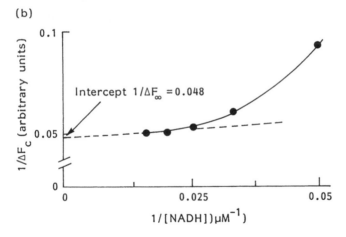

FIGURE 9.3: *Data from* Figure 9.2 *plotted (a) as ΔFc vs. [NADH], to show fluorescence change on binding and (b) as inverse plot to estimate ΔF$_\infty$ = 20.8 arbitrary units.*

(iii) The difference curve can then be calibrated in units of fractional saturation (*Figure 9.3a,* right-hand axis). The dissociation constant for NADH (K_d – a low K_d indicating a high affinity) can then be estimated from this curve. Mathematical analysis of the binding isotherm shows K_d smaller than or equal to [NADH] at 50% saturation (K_d = NADH at 50% saturation when NADH >> [protein] in the titration. If [NADH] ≈ [protein], K_d will have a lower value). Analysis of the data to obtain a precise value for K_d in the latter case is given in standard physical biochemistry texts [see Bagshaw and Harris (1987) *Spectrophotometry and*

Spectrofluorimetry: a Practical Approach (D.A. Harris and C.L. Bashford, eds), pp. 91–113. IRL Press, Oxford].

9.3 Measurement of the rate of ligand binding to a protein

9.3.1 Considerations of instrumental response time

A small molecule will bind to a protein at about the same rate as the two collide in solution. This means that, at reasonable concentrations of protein and ligand, a binding equilibrium will be established within 1 sec, that is, within the mixing time when conventional stirring is used to mix the reagents. To measure such rates, therefore, we need a rapid mixing device and a rapidly recording instrument (see Chapter 4).

Since spectrophotometry and fluorometry generate electrical signals, they are in principle ideal techniques for rapid measurement. The signals can be generated without manually sampling the reaction vessel, and transmitted instantaneously to the recording device. Some instrumental characteristics, however, may limit the speed of response of a particular instrument.

(i)   The response time may be limited by the mechanical movement of a chopper in a split beam or dual wavelength instrument, or by the pulsation time with a pulsed light source (see Chapter 4). Clearly, the instrument cannot record the rate of processes in which significant changes occur during the period when the measuring beam is interrupted. For rapid reaction studies, a continuous (nonpulsed) light source is generally required. In most modern instruments, the chopper moves rapidly enough to allow measurement in the millisecond time range, although this should be checked in the specifications of the instrument.

(ii) The response time of the measuring device may limit the time resolution of the instrument. While electronic devices, such as photomultipliers, do respond instantaneously to changes in light intensity, it is customary in laboratory instruments to sample the signal by sequentially averaging over a finite time, say 0.1 sec, to reduce instrumental noise. This is sometimes known as 'damping' the response. Clearly an instrument for measuring rapid reactions must have access to much shorter averaging times (1 msec approx.) without unacceptably increasing noise levels. This again should be checked in the manufacturer's specifications. A convenient rule of thumb is that the response time of the instrument should be less than 20% of the time constant of the process under study.

With a modern spectrophotometer or fluorometer, these factors limiting time resolution are readily eliminated by an appropriate choice of instrument, and appropriate choice of settings of the instrument. More problematic is the requirement to initiate the reaction (mix the reagents) within a few milliseconds. Conventional stirring devices are no use over this time scale; it is necessary to use a rapid mixing accessory which can be linked into the spectrophotometer in question.

9.3.2 A rapid mixing device

For spectrophotometric and fluorometric studies, a stopped-flow device is convenient for initiating rapid processes. A commercial device to fit an existing spectrophotometer is shown in *Figure 9.4a*. Its structure is shown diagrammatically in *Figure 9.4b*. The two solutions to be mixed are forced from two hydraulically driven syringes into a tangential mixing chamber. This induces turbulence which causes the solutions to mix rapidly. After mixing, the solutions flow into an observation chamber and then out into a stopping syringe, whose plunger is stopped by a metal block after a suitable amount of liquid has entered the observation chamber. Measurement is initiated just before the flow is stopped using a trigger switch; after the flow is stopped, the reaction mixture ages in the observation chamber and the reaction can be followed.

This device is set up in (or around) the sample compartment of a spectrophotometer, with the observation chamber replacing the conventional cuvette in the light path. For absorbance measurements, the long axis of the cell lies along the light path, to maximize l through the sample (thus maximizing the observed signal, as $A = \varepsilon.c.l$) while minimizing the volume of reagent required. For fluorescence measurements, a linear flow-through cell perpendicular to the light beam (typically square in cross section with 2 mm × 2 mm internal dimensions) is more convenient.

9.3.3 Example 2: Measuring the rate of ADP binding to myosin subfragment 1 (S1)

Experimental method.

(i)   Myosin S1 is the ATP-hydrolyzing head of myosin, from muscle. Like many proteins, it contains tryptophan, which fluoresces at 345 nm when excited at 280 nm. ADP is not fluorescent.

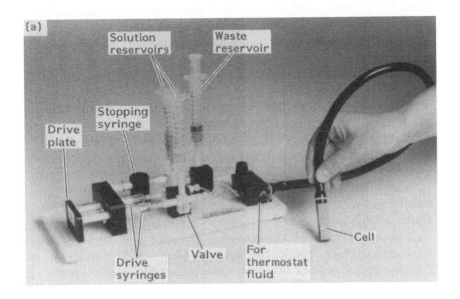

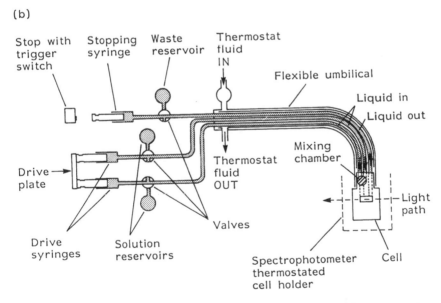

FIGURE 9.4: *Simple stopped-flow attachment for use with a conventional spectrophotometer (Hi-Tech SFA-20). For explanation see Section 9.3.2. Photograph reproduced by courtesy of Hi-Tech Ltd, Salisbury, UK.*

(ii) Myosin S1 (0.2 μM) is placed in a conventional fluorometer cuvette and its emission spectrum measured (λ_{ex} = 280 nm). ADP (20 μM) is then added, the solution mixed, and the spectrum determined again. After addition of ADP, fluorescence is higher (possibly because ADP, when bound, prevents solvent quenching of tryptophan – see Chapter 2). This suggests that ADP binding to myosin S1 can be monitored by fluorometry (λ_{ex} = 280 nm, λ_{em} = 345 nm).

(iii) Myosin S1 (0.2 µM) and ADP (20 µM) are loaded into the two syringes of a stopped-flow mixing device (above) attached to a fluorometer set to the above wavelengths. The connecting tubes are flushed out with the two solutions.

(iv) Binding is initiated by rapidly driving the two syringes forward, and (continuous) measurement begun just before the flow stops. The results are shown in *Figure 9.5a*. The trace consists of three phases – an initial unchanging fluorescence (representing the small amount of reaction that has occurred during flow between the mixing chamber and the observation point), a rapid and nonlinear increase in fluorescence (occurring in the stopped solution as ADP binds to the protein), and a final static fluorescence (representing the situation when the process is complete).

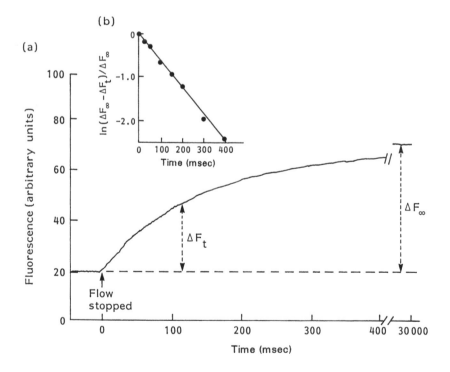

FIGURE 9.5: *Time course of tryptophan fluorescence change as ADP binds to myosin Sl. (a) Increase in tryptophan fluorescence with time. (b) Semi-logarithmic plot of the data in (a). For experimental conditions see Section 9.3.3.*

Calculation of rate constant. (*NB.* Conditions for this experiment were chosen so that (a) the process of ADP binding to myosin is pseudo-first order ([ADP]>>[myosin]), and (b) the process reaches 'completion' ([myosin]$_{free}$ ≈ 0 at infinite time). This greatly simplifies the mathematical analysis, but it should be noted that the experiment is not limited to these conditions, providing an appropriate mathematical treatment is used.

(i) The fluorescence change during the second phase represents the process of ADP binding to myosin. However, the observed change is nonlinear in time showing that the rate of ADP binding is continuously changing (as free myosin is being consumed). The appropriate parameter to describe the binding process is the second order rate constant, k (M^{-1} sec^{-1}), where k is defined as

rate of binding = $d[M \cdot ADP]/dt = -d[M]/dt = k[ADP][M]$.   (9.1)

To calculate k we need to manipulate the data collected.

(ii) Since this section of the graph represents the variation of [M·ADP] with time, we convert Equation 9.1, which relates rate to concentrations of reactants, to Equation 9.2, which relates concentrations to time. This involves integrating Equation 9.1.

$$\int_{[M] = [M]_o}^{[M] = [M]_t} -\frac{d[M]}{[M]} = k[ADP] \int_{t = 0}^{t = t} dt$$

$$ln\frac{[M]_t}{[M]_o} = -k[ADP]t \tag{9.2}$$

(Note that [ADP]>>[M]$_o$, so that [ADP] can be treated as a constant in the integration.)

(iii) Knowing that 1 mol ADP binds to 1 mol myosin S1, we can assume that the observed change in fluorescence ΔF (see *Figure 9.5a*) is proportional to the amount of ADP bound.

$$[M \cdot ADP]_t = \kappa(\Delta F)_t$$

where κ is the empirical proportionality constant (see Chapter 1, Section 1.7). The concentration of unreacted myosin ([M]$_t$) is correspondingly given by $\kappa (\Delta F_\infty - \Delta F_t)$, and the amount of free myosin present initially ([M]$_o$) by $\kappa (\Delta F_\infty)$.

(iv) Substituting these concentrations into Equation 9.2, we have

$$ln\frac{(\Delta F_\infty - \Delta F_t)}{(\Delta F_\infty)} = -k[\text{ADP}]t$$

and the proportionality factor, κ, need not be determined.

(v)   A plot is made of $ln\dfrac{(\Delta F_\infty - \Delta F_t)}{(\Delta F_\infty)}$ versus t (*Figure 9.5b*), and the

slope, $k[\text{ADP}]$ is determined. Since [ADP] is known, k can be calculated. The experiment is repeated at other values of [ADP]. and the values of k obtained are averaged.

(vi)   Note that, if the plot in *Figure 9.5b* is not a straight line, processes other than the postulated simple binding reaction must be occurring. Often this implies a conformational change in the protein subsequent to binding.

9.4 Measuring the concentration of ions inside cell compartments

---

A living cell is not simply a bag of chemicals of uniform concentration throughout. Animal and plant cells, in particular, contain cytoplasm bounded by a plasma membrane, but within this cytoplasm there are a number of smaller compartments (nuclei, mitochondria, vacuoles, endoplasmic reticulum, etc.), each enclosed by its own membrane. Just as the cytoplasm will have a different composition from the extracellular medium, these compartments will have different compositions from the cytoplasm, the composition reflecting the function of the compartment. Lysosomes, for example, form digestive vesicles inside the cell; their enzymes work at a pH optimum of around pH 5.5 and hence their internal H^+ concentration is considerably higher (pH lower) than that of the cytoplasm (pH 7.1). The determination of the concentrations of chemicals within different cell compartments is thus a step towards understanding the functions of such organelles in the cell.

Furthermore, many cell processes are controlled by the movement of ions between cell compartments. For example, the release of Ca^{2+} ions from the sarcoplasmic reticulum into the cytoplasm triggers muscle contraction. To understand such processes, therefore, we need to be able to follow the movement of ions from one cell compartment to another.

To determine the ionic composition in the various cell compartments is not simple. Classical biochemical techniques, involving fractionating the cell into its components, separating them, and then analyzing them, is slow and (unsurprisingly) suffers from a redistribution of ions during cell breakage or during the separation process. Classical physiological techniques, involving inserting an ion-specific electrode into the compartment under study, are similarly problematic as (i) the intracellular organelles are normally too small for even the finest electrode tip, and (ii) insertion of an electrode will itself damage membranes and cause ions to redistribute. Spectrophotometric and fluorometric measurements are rapid and noninvasive and are thus ideal in principle for measuring intracellular ion distributions, provided, of course, that a suitable chromophoric indicator can be located in the appropriate cell compartment. Suitable indicators have been shown in Chapter 2; much spectacular chemistry has been involved in their design, but its discussion lies outside the scope of this book.

As noted above, one benefit of spectrophotometric measurements is that they do not damage the cell, and this, combined with their rapidity, means that they can be used to follow intracellular ion movements continuously over time. There is also a less obvious, but equally important feature of the use of chromophoric indicators for intracellular measurements. Enzymes, receptors and other proteins respond in the cell to the concentration of ions in solution (the 'free' concentration). They are unaffected by ions bound to other cellular components, or in insoluble precipitates, as clearly they cannot be reached by these ions and thus cannot be influenced by them. Chemical analysis of cell components, for example, will by its very nature yield the total ('bound' + 'free') concentration of each individual ion present irrespective of whether these ions are biologically active in influencing enzymes etc. This effect is very marked in the case of Ca^{2+} ions, which are important in regulating a wide range of cell processes. The free concentration of $[Ca^{2+}]$ in a cell is about 10^{-7} M, rising to a maximum of 10^{-6} M in a fully stimulated cell (e.g. actively contracting muscle). In contrast, the total $[Ca^{2+}]$ in tissues, as shown by chemical analysis, reaches 10^{-3} M. Thus only indicator methods, which can follow changes in free $[Ca^{2+}]$ between 10^{-7} and 10^{-6} M, are useful for investigating the mechanism and control of Ca^{2+}-triggered processes inside cells.

9.5 Types of optical probe

Environmental probes in biological systems may be classified in a number of ways. A useful distinction is between intrinsic and extrinsic probes. Intrinsic probes are naturally part of the system under study, like tryptophan in proteins, NADH as a dehydrogenase substrate (see Chapter 8, Section 8.7) or chlorophyll and carotenoids in photosynthetic systems (to be discussed below). Extrinsic probes, in contrast, are artificial compounds, designed and introduced by the experimentalist to report on particular aspects of the system. Examples are the acridines used for following pH changes and DNA structure, fura-2 for measuring intracellular [Ca^{2+}], and fluorescent ATP analogs for probing ATP binding sites (see Chapter 2).

Intrinsic probes clearly do not perturb the system under study, and are thus preferable where they are available. However, many biological systems do not contain appropriate chromophores or fluorophores, and even when they do, superior optical properties (higher quantum yield, higher sensitivity to environmental changes) and/or specificity for the change under study may favor the use of an extrinsic probe. In these cases, examples of which are given below, measurements of biological function are essential to establish that the system under study is not killed or irretrievably damaged by the added probe.

Another classification of probes is based on their physical behavior in use. In many cases, the probe is in a fixed position (within a membrane, at a binding site, in the cell cytoplasm, etc.) where it remains. Measurement in this case records signals corresponding to changes in the environment of this fixed position (changes in ion concentration at this point, potential gradient across the probed membrane, etc.).

In some cases, however, the probe responds to environmental changes by moving towards the environment it prefers. Weak bases, for example, tend to move towards acidic environments, and positively charged ions towards negatively charged regions. This is very noticeable in vesicular membrane systems, where a probe may cross the membrane and accumulate inside vesicles in response to a pH or electric potential gradient. Such probes are termed distribution probes – it is their movement to a new environment that is signalled to the observer (*Figure 2.11*).

Distribution probes are less suitable for measuring the kinetics of changes, since the rate of change of signal may reflect the rate of probe movement rather than the rate of change in the compartment studied. None the less, distribution probes have proved extremely useful in following changes in membrane systems. Examples of both types are given below.

9.5.1 Example 3: Measurement of intracellular [Ca²⁺] with fura-2 using ratio fluorometry

Theory. Fura-2 is a fluorescent Ca^{2+} chelator whose fluorescent properties change on binding Ca^{2+} ions. Its structure is shown in *Figure 2.12*. The affinity of this compound for Ca^{2+} is such that it binds Ca^{2+} well over the physiological range (K_d = 225 nM, while cytoplasmic [Ca^{2+}] = 50–2000 nM. It is thus a useful fluorescent indicator of free [Ca^{2+}] inside cells. It is also sufficiently fluorescent to be detectable at low concentrations (μM) and so does not itself perturb the cell by sequestering large amounts of Ca^{2+}.

Unusually, it is the excitation spectrum of fura-2 that changes on Ca^{2+} binding. While the emission peak remains fixed at around 510 nm, the excitation peak shifts from 380 nm (at [Ca^{2+}] = 0) to 340 nm (at [Ca^{2+}] = ∞) on binding Ca^{2+} (*Figure 4.10*). Thus fluorescence measured at 510 nm while exciting at 340 nm will increase with increased [Ca^{2+}] in its environment.

The measured level of fura-2 fluorescence will thus be proportional to (i) the concentration of Ca^{2+} ions complexed, and (ii) the amount of fura-2 in the light path. For intracellular measurements, however, the concentration of fura-2 inside the cell is not known. Even worse, its concentration may vary in time as fura-2 is destroyed by intracellular metabolism. Thus, intracellular [Ca^{2+}] is more reliably measured with fura-2 using ratio fluorometry, in which the fluorescence from excitation at 340 nm (fura-Ca^{2+}) is compared with that from excitation at 380 nm (free fura). This ratio of the concentrations of these species (and hence the fluorescence ratio) depends only on [Ca^{2+}] and not on the absolute concentration of fura present. This is particularly useful if an inhomogenous sample (e.g. a monolayer of cells) is to be investigated.

Measurement of the fluorescence ratio requires a dual wavelength (ratio) fluorometer as described in Chapter 4, Section 4.8. This possesses a single emission monochromator but two excitation monochromators, the sample receiving light from each alternately (at about 500 Hz) by means of a mechanical chopper.

Experimental method.

(i) Rat ventricular cardiomyocytes may be cultured in free suspension or in a monolayer. If in free suspension, the suspension itself may be studied in a standard fluorescence cuvette. If in a monolayer, this should be attached on a glass or plastic sheet of suitable dimensions to fit diagonally across the cuvette.

(ii) Fura-2 is introduced into the cell cytoplasm as its acetoxymethyl ester (Chapter 2). The ester is initially dissolved at high concentration in dry dimethylsulfoxide and mixed with a mild surfactant (e.g. Pluronic F127) and albumin to increase its solubility in the cell suspension medium. A small volume of this concentrated solution is then introduced into the medium. The nonpolar ester crosses the cell membrane and enters the cytoplasm where it is hydrolyzed to the carboxylic acid (fura-2 itself). This is charged and membrane-impermeant, so it remains trapped in the cytoplasmic compartment. After about 30 min with occasional mixing, external dye/ester are washed away, directly by flushing with medium in the case of monolayers or by gentle centrifugation and resuspension in the case of free cells.

(iii) The cells loaded with fura-2 are then introduced into the fluorometer, thermostatted at 37°C, and their fluorescence emission measured at 510 nm using excitation wavelengths chopped between 340 nm and 380 nm.

(iv) Manipulations of the cells (addition of hormones, drugs, electrical stimulation, etc.) can be carried out in the cuvette, and their effects on fluorescence determined. *Figure 9.6* shows the effect of electrical stimulation on [Ca^{2+}] levels in rat ventricular cardiac muscle cells; a spike of Ca^{2+} is released into the cytoplasm on electrical stimulation, and the level is rapidly restored to normal by cellular ion pumps. Isoproterenol, an analog of adrenaline increases the size, but not the duration, of this spike.

(v) As a blank control, the experiment is repeated with a control of untreated cells (not loaded with fura-2). Fluorescence observed after excitation at 340 nm and 380 nm ('autofluorescence') is subtracted from the experimental values above to obtain the actual signal due to fura.

Calibration and calculations.

(i) As has been seen in Section 2.7, the F_{340}/F_{380} ratio is clearly a qualitative measure of [Ca^{2+}] within the cardiomyocytes, its increase representing an increase in free intracellular [Ca^{2+}]. To determine quantitative values of intracellular [Ca^{2+}], we need to

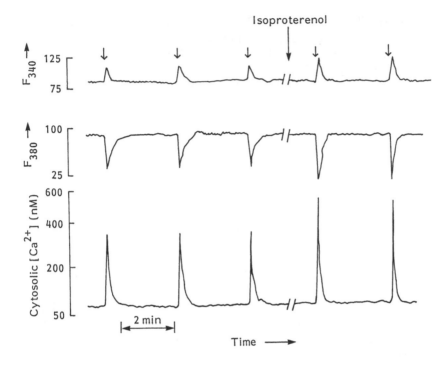

FIGURE 9.6: *Fluorescence changes of fura-2 inside a suspension of rat ventricular cardiomyocytes. Electrical stimulation is indicated by the bold arrows,and addition of isoproterenol by the thin arrow. The lowest trace indicates the concentration of free cytosolic Ca^{2+}, as calculated from the $\dfrac{F_{340}}{F_{360}}$ ratio (see Figure 9.7). Redrawn from McCormack and Cobbold (eds)*

(1991) Cellular Calcium: a Practical Approach, p. 42, by permission of Oxford University Press.

    determine the F_{340}/F_{380} ratio for known Ca^{2+} concentrations, that is, to calibrate the system. Owing to variations in lamp intensity at the two wavelengths (see Chapter 3), calibration needs to be carried out individually for the instrument in use.

(ii)   Calibration is conveniently carried out on the same cells loaded with fura-2, after the experimental manipulations have been made (see above). An ionophore, ionomycin, is added to the cells to make them permeable to Ca^{2+} ions. The external medium is then depleted of Ca^{2+} by adding the chelator, EGTA, and this in turn leads to removal of intracellular Ca^{2+} (via the ionophore). F_{340}/F_{380} at zero internal $[Ca^{2+}]$ can then be measured. A high concentration of $CaCl_2$ is then added to the medium; this enters the cells via the ionophore and saturates fura-2, allowing the measurement of F_{340}/F_{380} for the fura-Ca complex. Finally, $MnCl_2$

is added. This quenches all fluorescence due to fura, and allows the autofluorescence of the cells to be measured (*Figure 9.7*).

(iii) By definition, for the binding of Ca^{2+} to fura

$$K_d = \frac{[\text{fura}]\,[Ca^{2+}]}{[\text{fura} - \text{Ca}]} \quad \text{where } K_d \text{ is the (known) dissociation}$$

constant.

Thus $\left[Ca^{2+}\right] = K_d . \dfrac{[\text{fura} - \text{Ca}]}{[\text{fura}]}$

Mathematical manipulation shows that this converts to

$$\left[Ca^{2+}\right] = K_d \frac{(R - R_{min})}{(R_{max} - R)} . \frac{F'_{380}}{F''_{380}}$$

where $R = F_{340}/F_{380}$, $R_{min} = R$ when $[Ca^{2+}] = 0$, $R_{max} = R$ when $[Ca^{2+}]$ is saturating, F'_{380} = fluorescence of free fura when excited at 380 nm, and F''_{380} = fluorescence of fura-Ca when excited at 380 nm. Note that this gives a fixed, but nonlinear, relationship between the measured fluorescence ratio and free $[Ca^{2+}]$ (*Figure 9.7b*).

Variations on a theme.

(i) Changes in cytoplasmic pH can be measured by an analogous procedure using the fluorescent H^+ indicator (again trapped inside the cell by hydrolysis of its ester) 2',7'-*bis*-(2-carboxyethyl)-5 carboxyfluorescein (BCECF).

(ii) Changes in intramitochondrial $[Ca^{2+}]$ can be measured using fura-2 as above, provided that the fluorescence of cytoplasmic fura is quenched by adding $MnCl_2$ to the cells before the fluorescence measurements are taken. In the absence of a signal due to cytoplasmic fura, the (smaller) signal due to fura that has entered mitochondria can be detected.

9.6 Measurement of intravesicular pH using a distribution probe

9.6.1 Acridine dyes

Acridine dyes, which are large aromatic heterocyclic bases, are fluorescent molecules useful as environmental probes in a variety of systems. Acridine orange, for example, will give an enhanced

(a)

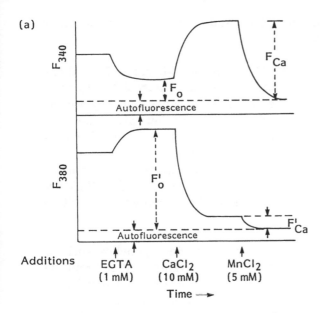

(b)

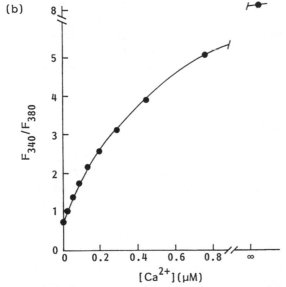

FIGURE 9.7: *(a) Determination of* R_{max} *and* R_{min} *for fura-2 fluorescence in heart cells. The cells were loaded with fura-2 and treated with ionomycin. Other additions were as indicated.* $R_{max} = \dfrac{F_{Ca}}{F'_{Ca}}$ $R_{min} = \dfrac{F_0}{F'_0}$ *. For explanation see Section 9.5.1. (b) Variation in* $\dfrac{F_{340}}{F_{380}}$ *ratio of fura-2 with increasing* $[Ca^{2+}]$.

Reproduced from McCormack and Cobbold (eds) (1991) Cellular Calcium: a Practical Approach, *p. 45, by permission of Oxford University Press.*

fluorescence when bound to DNA, and is used for intracellular staining of this nucleic acid.

Certain acridines, such as 9-AA and 9-amino-6 chloro-2 methoxy acridine (ACMA) are water soluble bases (see *Figure 2.11*). In their uncharged forms they can permeate cell membranes, but they become charged (and impermeant) on protonation at low pH. Thus if a membrane separates high pH and low pH compartments, 9-AA (or ACMA), in its protonated form, will accumulate on the low pH side (*Figure 2.11*). Both protonation and high concentration will quench acridine fluorescence (see Chapter 2). Thus 9-AA and ACMA are distribution probes, which will move in a pH gradient with a resulting decrease in fluorescence.

This is useful in studying energy coupling in organelles such as chloroplasts. On illumination, or when hydrolyzing ATP, chloroplasts accumulate protons into an internal space, the thylakoid vesicle; the resultant pH gradient is an important intermediate in energy coupling. In the example given, 9-AA is used to monitor the transmembrane pH gradient formed when isolated chloroplasts are illuminated. It also exemplifies the problems encountered when the system under study is activated by light (the actinic beam), at the same time as a photometric measurement (using a measuring light beam) is carried out.

9.6.2 General considerations

(i)   9-AA will fluoresce with λ_{max} for excitation at 360 nm and λ_{max} for emission at 490 nm. Since a pH gradient leads to fluorescence quenching (fluorescence falls at 490 nm), measurements can be made with a single excitation and single emission monochromator, that is, on a standard right-angle fluorometer. However, since chloroplasts are to be studied under illumination (which drives electron flow), commercial fluorometers require modification to allow illumination by the actinic beam as well as by the measuring beam. As a corollary, the measuring beam should be of low intensity (i.e. the excitation slit kept to a minimum) so that it does not, itself, drive significant photosynthetic electron transfer.

(ii)   It is important that the photomultiplier measuring fluorescence does not record light from the actinic beam, either directly or by scattering. If it did, as this beam is relatively intense, any changes in the acridine fluorescence would be observed over a large background signal of unchanging illumination. This problem is overcome by: (i) orienting the actinic light beam orthogonal to the plane of the measuring light beam and photomultiplier (*Figure 9.8a*); (ii) filtering the actinic light beam so as to exclude wavelengths involved in fluorescence measurement, if possible; and (iii) in some instruments, using a pulsed light source (with a lock-in amplifier) for the measuring beam. Here the amplifier will record only modulated light, and it hence 'ignores' the continuous actinic illumination.

9.6.3 Experimental method

(i)   A standard right-angle fluorometer is modified by drilling a hole in the lid of the sample compartment vertically above the cuvette holder. (In some instruments, such as the Perkin-Elmer LS series, the sample compartments are readily interchangeable (*Figure 9.8a*), obviating the need for a dedicated instrument.) This hole is fitted with a fiber optic to a light source, which is filtered to block wavelengths below 600 nm. An injection port may also be drilled through the lid (*Figure 9.8a*).

(ii) The cuvette holder is thermostatted, and a stirred cuvette containing buffer is added and allowed to reach a suitable temperature (25°C). Chloroplasts are added to a final concentration of 20 µg chlorophyll m^{-1} and 9-AA to 20 µM. (Higher 9-AA concentrations will uncouple the chloroplasts.)

(iii) Fluorescence (λ_{ex} = 360 nm, λ_{em} = 490 nm) is measured. The actinic beam is then switched on, and fluorescence is followed until a new steady state is reached (about 30 sec). Switching off the actinic beam allows the fluorescence to return the initial (high) level (*Figure 9.8b*).

(iv) Various effectors (mediators or inhibitors of electron flow, herbicides, uncouplers, etc.) can be added through the injection port and their effects on the chloroplasts followed.

(v)   As a control, the experiment is repeated in the presence of chloroplasts but in the absence of 9-AA. If the actinic beam has been adequately filtered, the signal should be less than 5% of that seen in the presence of 9-AA, and unchanged when the actinic beam is switched on or off.

9.6.4 Calculations

(i)   9-AA will ionize in solution, to an extent determined by its acid dissociation constant.

$$K_a = \frac{[A][H^+]}{[AH^+]} \qquad (9.3)$$

(a)

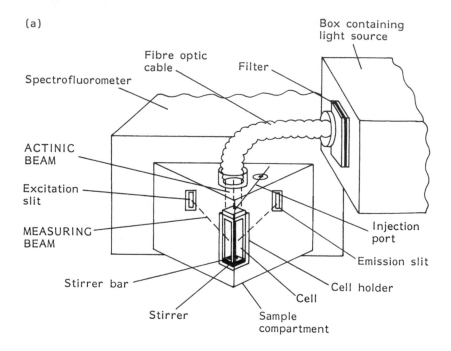

(b)

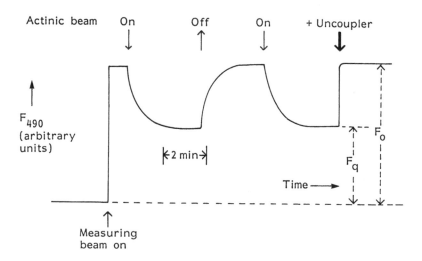

(ii)   On one side of the membrane (say, the inside)

$$K_a = \frac{[A]_{in}[H^+]_{in}}{[AH^+]_{in}} \text{ and, similarly } K_a = \frac{[A]_{out}[H^+]_{out}}{[AH^+]_{out}}$$

However, $[A]_{in} = [A]_{out}$, since the unionized form, A, can freely permeate across the membranes, Thus, since K_a is a constant inside and outside the vesicles

$$\frac{[H^+]_{in}}{[H^+]_{out}} = \frac{[AH^+]_{in}}{[AH^+]_{out}} \tag{9.4}$$

(iii) Taking logs, and using the approximation that virtually all the acridine is protonated at the pH of study (approximately pH 7), we have the equation

$$\Delta pH = \log\frac{[A_t]_{out}}{[A_t]_{in}} \tag{9.5}$$

where $[A_t]$ = total concentration of 9-AA (= protonated + unprotonated) in the relevant compartment.

(iv)   The fluorescence of 9-AA is essentially completely quenched when it is taken up into the vesicle interior. Thus, if quenching is given by

$$Q = \frac{F_o - F_q}{F_o}$$

(where F_o is the fluorescence in the absence of any pH gradient, and F_q in its presence), the amount of 9-AA taken up into the vesicles is proportional to the quenching observed.

FIGURE 9.8: *Measurements of 9-AA fluorescence in photosynthetic vesicles. (a) Instrumental set up. (In the Perkin-Elmer LS series depicted, the sample compartment takes the shape of a triangular prism, bolted on to the front of the instrument.) Light from a stabilized light source is passed through a suitable filter (to select wavelengths to drive photosynthesis) and is led, through a fiber optic cable, through the lid of the sample compartment to illuminate the cuvette from above. The measuring beam (which is modulated to avoid interference from the constant actinic beam) enters the sample compartment from the fluorometer through the excitation slit and fluorescence is measured at right angles as light re-entering the fluorometer via the emission slit. An injection port and stirrer allows additions to be made to the sample. (b) Response of 9-AA in chloroplast suspension on illumination or addition of uncoupler. F_o is the fluorescence observed in the absence of actinic illumination, F_q when illumination is present. For explanation see Section 9.6.3.*

namely $$[A_t]_{(in)} \propto]Q$$

The concentration of 9-AA now inside the vesicles is also related to the volumes inside (v) and outside (V)

namely $$[A_t]_{in} \propto Q\frac{V}{v}$$

Similarly,

$$[A_t]_{out} \propto (1 - Q)$$

Therefore $$\frac{[A_t]_{in}}{[A_t]_{out}} = \frac{Q}{1 - Q} \cdot \frac{V}{v} \qquad (9.6)$$

or, substituting in Equation 9.3

$$-\Delta pH = \log\frac{[Q]}{[1 - Q]} + \log\frac{V}{v}$$

(where the negative sign indicates a lower pH inside the vesicles; v, the volume inside thylakoid vesicles is around 10 µl mg^{-1} chlorophyll; V is of course the volume in the cuvette.)

9.7 Measurement of transmembrane potential

Unequal movement of positively and negatively charged ions across biological membranes means that such membranes typically separate compartments of differing electrical potential. Thus the cytoplasm of a mammalian cell is typically 70 mV more negative than the extracellular fluid. (Conventionally, this is said to represent a membrane potential of –70 mV.) Since functioning mitochondria pump out positively charged (H$^+$) ions, their interior is a further 150 mV more negative.

These potentials are vital for cell functioning, being used to drive transport systems, ATP synthesis, and (in nerve cells) they are used for signalling. Monitoring them is thus of interest to the biochemist and physiologist. Except in very favorable circumstances (e.g. the giant axons of squid), direct monitoring of electrical potentials by electrodes is impractical in living cells; the electrode tip is too large to insert into cells (or organelles) without damaging them. Thus spectrophotometric methods are conveniently employed.

Various probes are available for monitoring membrane potential, and these may be classified as was discussed in the previous section. Thus oxonol (lipophilic anions) and cyanine dyes (lipophilic cations) (*Figure 6.10*) are extrinsic probes whose distribution across the membrane and/or bound to its surface varies with membrane potential. Their fluorescence changes with their distribution and, with suitable calibration, these probes allow membrane potential to be followed fluorimetrically essentially as described for the use of acridines in monitoring transmembrane pH above (Section 9.8).

An alternative method uses the carotenoids found in photosynthetic membranes as an intrinsic probe. These are long, conjugated compounds whose excited state dipole is high compared with the ground state and, as a result, their absorption spectrum shows a red shift when the molecule is in an electric field. This electrochromism is sometimes termed the Stark effect (Section 2.3).

Where carotenoids are present, this method has several advantages over methods using the cyanine or oxonol dyes. Firstly, as an intrinsic probe, carotenoids do not perturb the system under study. Secondly, since carotenoids are fixed within the membrane, their response is rapid (within μsec) and does not rely on the rate of probe movement through the membrane.

9.7.1 Example 4: Determination of light-induced membrane potential in chromatophores from a photosynthetic bacterium

Experimental method.

(i)   As the wavelength shift, and the corresponding change in absorbance, is relatively small ($\Delta A \leq 0.05$), and measurements are carried out on suspensions of membrane vesicles rather than clear solutions, the carotenoid shift is measured using a dual wavelength spectrophotometer (see Chapter 4). The sample compartment is adapted for side illumination with an actinic beam in addition to the measuring beam (see Section 9.6).

(ii) Chromatophores (10 μM bacteriochlorophyll) are added to a cuvette containing buffer. Using the instrument in dual beam mode, a difference spectrum is taken between illuminated and dark chromatophores in the region 450–550 nm, where the carotenoid absorbance is greatest (*Figure 9.9*). This establishes (a) the wavelength suitable for measuring the carotenoid shift (about 505 nm) and (b) a nearby reference wavelength (isosbestic point) (about 496 nm). The precise wavelengths will depend upon the species and strain of bacteria from which the chromatophores are derived.

(iii) The instrument is then changed to dual wavelength mode using these two wavelengths. (Note that, in *Figure 9.9a*, the measuring wavelength is at an absorbance trough; absorbance will decrease as membrane potential increases.) $\Delta A_{(505-493)}$ is then measured under the conditions required (continuous illumination (*Figure 9.9b*), short flashes, with reagents such as uncouplers, etc.).

(iv) For greater precision in measuring such small changes in absorbance, especially with brief flashes, computer accumulation of the data over short periods (1–5 sec) can be triggered by switching on the actinic beam. The experiment can then be repeated and the average of several accumulations taken as the experimental data (not shown).

Calibration and calculations.

(i) The carotenoid absorbance changes linearly with the magnitude of the applied field, here the membrane potential. To calibrate the system, a diffusion gradient of ions is set up artificially across the membrane, and the resulting ΔA measured. This is achieved as follows:

(ii) The chromatophores are suspended in a low K^+ medium (say 0.1 mM). Electron transfer is blocked in the membranes by the inhibitor, antimycin *A*, and the membrane made permeable to K^+ using the ionophore, valinomycin.

(iii) KCl is added to the cuvette, to a final K^+ concentration of, for example, 10 mM. As K^+ (but not Cl^-) can now cross the membrane, a diffusion potential $\Delta\Psi$(positive inside) is induced as K^+ moves in. The magnitude of this potential is given by the Nernst equation.

$$\Delta\Psi = \frac{RT}{F} \, ln \, \frac{[K^+]_{out}}{[K^+]_{in}}$$

ΔA is measured immediately after K^+ addition; it will decay slowly as anions leak into the vesicles.

FIGURE 9.9: *Measurement of membrane potential in photosynthetic membranes using the carotenoid shift. (a) (Illuminated-dark) difference spectrum of chromatophores from a photosynthetic bacterium in the region of carotenoid absorption. Isosbestic points are indicated by arrows. (b) Changes in A_{505} with energy state of the chromatophores. (c) Calibration of the response of the carotenoids ($\Delta A_{505-496}$) to a diffusion potential induced by a KCl gradient. For experimental details see Section 9.7.1. Data from Jackson and Crofts [(1971) FEBS Lett., 4, 185–189] and Ferguson et al. [(1979) Biochem. J., 180, 75–85].*

(a)

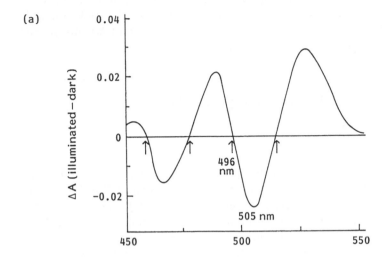

(b)

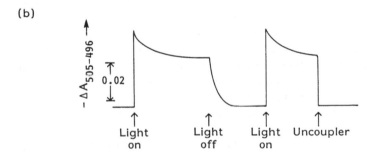

(c)

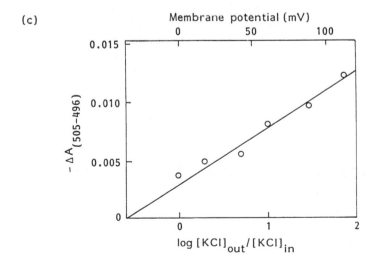

(iv) The measurement is repeated after the addition of various amounts of KCl. ΔA is then plotted against $\Delta\Psi$ (*Figure 9.9c*), and from this line, $\Delta\Psi$ can be read off for any value of ΔA measured in the experimental section above.

9.8 Fluorescence energy transfer as a molecular ruler

After absorbing a photon, a fluorophore may release its excited state energy as light, or allow it to be degraded to heat (Chapter 1). If, however, another chromophore is nearby, the energy may instead be transferred directly to this second molecule by resonance between their electronic transitions. This phenomenon is termed fluorescence resonance energy transfer (FRET), and does not involve the intermediary of a photon; energy is transferred by direct electronic interaction between the molecules (Chapter 2, Section 2.4). This transfer is observed as a decrease in quantum yield of the first (donor) fluorophore and, if the second chromophore is also a fluor, as fluorescence from this second (acceptor) chromophore. Emission from the acceptor will be at a higher wavelength than that expected from donor fluorescence, since a fluorophore always emits at a lower energy (higher wavelength) than the energy absorbed (Chapter 1, Section 1.6).

Energy transfer requires an overlap between the emission spectrum of the donor and the absorption spectrum of the acceptor (*Figure 2.8*). More particularly, the efficiency of energy transfer falls off sharply with the distance (R) between the donor and acceptor molecules, transfer dropping off as $1/R^6$. It is this relationship that allows FRET to be used in distance measurements as a molecular ruler. Typically, this method can be used to determine interchromophore distances of 2–10 nm. This is of the order of magnitude of the size of protein molecules. Such techniques have been used, for example, to map the proximity of different ribosomal proteins in the ribosome.

9.9 Probing the structure of a four-way DNA junction

9.9.1 Introduction

During recombination of DNA, in meiosis, DNA duplexes of homologous sequence align and exchange strands in a four-way DNA junction (a Holliday junction). This is shown in 2D representation in *Figure 9.10d*. A model of such a junction can be made using complementary oligonucleotides (*Figure 9.10a*), and this model used to probe the 3D conformations available to this structure.

The oligonucleotides form helical duplexes that are connected at a central point. The helices may lie parallel to each other, or cross at right angles or at some other angle. Furthermore, the sequences may run parallel (crossed configuration) (*Figure 9.10*) or antiparallel (uncrossed) (*Figure 9.10c*) These possibilities can be distinguished if the distances between all pairs of 5′ ends are known, and these measurements can be made by fluorescence energy transfer.

9.9.2 Experimental method

(i)   Oligodeoxyribonucleotides of 34 base length (34-mers) are synthesized chemically to a specified sequence (*Figure 9.10a*). An aminohexyl group is attached to the 5′ end, as a group for derivatization. This is then reacted with a fluorescent probe, either fluorescein isothiocyanate (the donor fluor) or tetramethyl rhodamine N-hydroxysuccinimide (the acceptor fluor) under conditions where the reaction is complete to 1 mol label per DNA strand. For each of the 4 complementary strands a fluorescein and rhodamine derivative is prepared.

(ii)   A fluorescein-labeled oligonucleotide is dissolved in buffer at about 10^{-7} M (absorbance <0.05, to minimize the inner filter effect). Its fluorescence emission spectrum (λ_{ex} = 490 nm) and excitation spectrum (λ_{em} = 520 nm) is recorded using conventional right angle optics. This is repeated for a rhodamine labeled oligonucleotide (λ_{ex} = 565 nm, λ_{em} = 590 nm). These spectra provide measurements in the absence of energy transfer. They allow selection of an excitation wavelength which will mainly excite the fluorescein donor (λ_{ex} = 490 nm) and an emission wavelength (λ_{em} = 600 nm) where fluorescein fluorescence is minimal but rhodamine fluoresces well.

(iii) The four oligonucleotides, including one (say P) with a fluorescein-labeled 5′-end, and one (say Q) with a rhodamine-labeled 5′-end are allowed to anneal to form the four-way junction. An emission spectrum is then measured, using λ_{ex} = 490 nm. If energy transfer occurs, this spectrum shows a peak of rhodamine fluorescence in addition to one due to fluorescein (i.e., light absorbed by the fluorescein has caused the rhodamine to fluoresce) (*Figure 9.11*). An excitation spectrum (λ_{em} = 600 nm) is also measured for the complex.

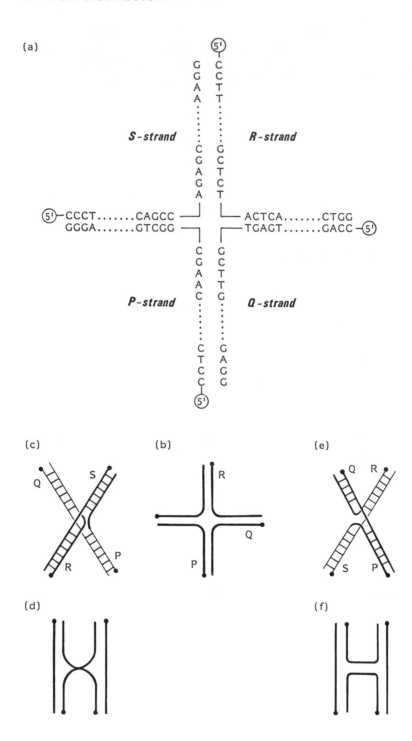

This experiment is repeated with different pairwise combinations of labeled oligonucleotides (P–Q, P–R, P–S, Q–S, R–S, Q–R) so that all six distances between the four ends can be determined (*Figure 9.11*, inset).

(iv) The effects of various conditions on the structure of the complex can be studied. By synthesizing different oligonucleotides, the effects of oligonucleotide length, base sequence, or of unpaired bases can be investigated. Also important are the effects of ions; for any significant energy transfer, physiological conditions of $[Mg^{2+}]$ (5 mM) must be present to neutralize the negatively charged phosphate groups and thus allow the helices to pack together.

9.9.3 Calculations

(i) With the oligonucleotides given, fluorescence energy transfer is observed only with the labeled pairs P–S and Q–R (*Figure 9.11*, inset), indicating that the structure places their 5′ ends closer in space than occurs with any other possible pairs. (Note: this orientation is sensitive to base sequence in the central region of the junction; the replacement of two base pairs in the central region of the complex, brings the PQ and RS ends closer together.)

(ii) To obtain a precise measure of the interprobe distance, we need to calculate the efficiency of energy transfer, E.

$$E = \frac{\text{actual acceptor emission (excited via FRET)}}{\text{acceptor emission for 100\% transfer}} = \frac{R_o^6}{R^6 + R_o^6} \quad (9.7)$$

where R is the distance to be determined and R_o is a constant that depends on the spectral overlap between the two probes, the fluorescence lifetime of the donor and other physical constants. (One of the physical 'constants' relates to the relative orientation

---

FIGURE 9.10: *Arrangements of DNA strands at a four-way junction.* *(a) Synthetic nucleotides forming a four-way junction. The dots represent additional bases (paired) in the 34 base oligonucleotides. (b) Two-dimensional representation of this structure with all arms equidistant. (c,e) Representation of possible three-dimensional arrangements of these strands. (c) A 'crossed structure' with the two unbent strands running parallel. (e) A 'noncrossed' structure with the two unbent strands running antiparallel. Note that each of these can exist in a structure of opposite handedness (not shown). (d,f) Representation of (c) and (e) in terms of a 'Holliday junction'. Redrawn from Murchie et al. (1989) Nature, 341, 763–766 with permission from Macmillan Magazines Ltd.*

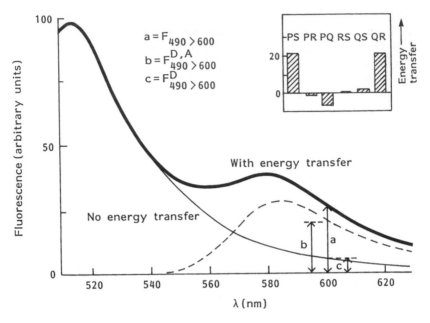

FIGURE 9.11: *Fluorescence energy transfer from fluorescein-labeled to rhodamine-labeled DNA strand. Excitation was 490 nm (λ_{max} for fluorescein). The upper curve shows fluorescence when energy transfer to rhodamine occurs, the lower, the emission spectrum of fluorescein alone.The difference curve (···) shows rhodamine fluorescence due to energy transfer. The values a, b and c are used in the calculation of distance (see Section 9.9.3). $F_{490>600}$ being negligible under these conditions. (Insert) Energy transfer, as indicated by the ratio $\dfrac{F^{DA}_{490>600}}{F^{A}_{565>600}}$ see Section 9.9.2. Oligonucleotides are labeled at the 5′ end of the strands indicated. Data from Murchie et al. [(1989) Nature, **341**, 763–766] and Clegg [(1992) Methods Enzymol. **211**, 353–388].*

of the two probes, and may not be precisely constant in all systems.) It can be found in the literature, for the fluorescein–rhodamine pair, R_o = 4 nm. (From Equation 9.7, above, R_o can be seen to be the distance apart of the two chromophores for the efficiency of transfer to be 50%.)

(iii) The actual emission from the acceptor due to FRET is the emission at 600 nm ($F_{490>600}$) minus the contribution from fluorescein fluorescence at 600 nm with λ_{ex} = 490 nm ($F^D_{490>600}$) and from rhodamine excited directly at 490 nm ($F^A_{490>600}$) (*Figure 9.11*). These latter contributions should be quite small, and can be measured from the emission spectrum (λ_{ex} = 490 nm) of each probe taken individually. Thus the true emission due to FRET , $F^{D,A,}$ is given by:

$$F^{D,A}_{490>600} = F_{490>600} - F^{D}_{490>600} - F^{A}_{490>600}$$

(iv) The expected emission for 100% efficiency of transfer is estimated from the fluorescence of rhodamine excited at a wavelength where fluorescein does not absorb ($F^{A}_{565>600}$), which can be measured from the excitation spectrum of the complex (λ_{em} = 600 nm) corrected for the relative absorbances of the donor (fluorescein) at 490 nm compared with the acceptor (rhodamine) at 565 nm.

$$F_{100\%} = F^{A}_{565>600} \cdot \frac{\varepsilon^{D}_{490}}{\varepsilon^{A}_{565}}$$

(This expression assumes equal concentrations of donor and acceptor fluorophores. This would be the case at probe/DNA molar ratios of 1.) Measuring F^{A} and $F^{D,A}$ in the same cuvette under the same conditions, as here, obviates any problems due to the dependence of measured fluorescence on geometry, amplifier gain, etc.

(v) Thus

$$E = \left(\frac{F^{D,A}_{490>600}}{F^{A}_{565>600}} \right) \cdot \frac{\varepsilon^{D}_{490}}{\varepsilon^{A}_{565}} = \left(\frac{R_o^6}{R^6 + R_o^6} \right)$$

R_o, ε_{490} and ε_{565} are known values from the literature; measurement of $F^{DA}_{490>600}$ and $F^{A}_{565>600}$ thus lead to the determination of R. In this case, $R \approx 6$ nm between the closest ends, suggesting an angle of about 60° between the arms. These data are consistent only with an 'uncrossed' model with the helices stacked at an angle of about 60°, one strand in the major groove of the other helix. Building a molecular model of this structure leads to the model as shown in *Figure 9.12*.

(a)

(b)

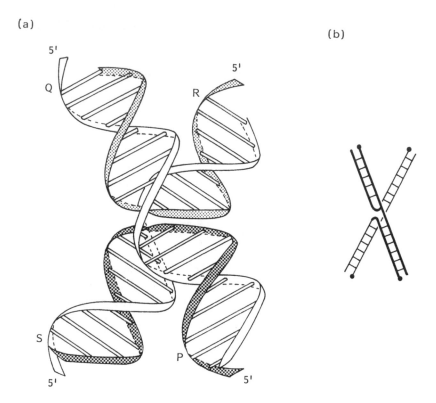

FIGURE 9.12: Model of DNA 4-way junction derived from fluorescence energy transfer measurements. (a) Three-dimensional representation. (b) Schematic (5′ termini represented by filled circles helices at the rear are stippled. Reproduced from Murchie et al. (1989) Nature, **341,** pp. 764 and 765, with permission from MacMillan Magazines Ltd.

Appendix A

Glossary

Absorbance (A_λ): quantitative measure of absorption at wavelength λ, defined as log (I_o/I), where I_o is the intensity of the incident light, and I, the intensity of the transmitted light at this wavelength.

Absorption: a decrease in light transmitted through a solution due to removal of energy by electronic excitation of molecules in the light path. The size of particles giving rise to absorption must be considerably less than the wavelength of illumination.

Actinic illumination: illumination driving a photochemical process (e.g. electron release from chlorophyll in photosynthesis), rather than that employed for measuring absorbance or fluorescence.

Action spectrum: dependence of a photochemical process on the wavelength of actinic illumination (q.v.).

Assay: quantitative determination of the amount of substance under study.

Beer-Lambert law: the linear relationship between absorbance and concentration/path length. $A = \varepsilon.c.l$.

Carotenoid shift: shift of absorption peak of the carotenoids of photosynthetic systems on actinic illumination. The shift occurs between 500–550 nm, depending on the organism studied, and is due to the effect of the electric field induced across the membrane on the stability of the excited state (cf. Stark effect).

Chopper: rapidly rotating mask or mirror which interrupts the beam illuminating the sample (or switches between two illuminating beams) at regular intervals.

Chromophore: that atomic grouping within a molecule, or system, responsible for the absorption of light.

Cuvette: transparent sided vessel (cell) for holding liquid samples in the optical path of a spectrophotometer or fluorometer.

Difference spectrum: the result of subtracting the spectrum of a compound from a spectrum of the same compound after modification by a chemical or physical transition. This can be determined directly in a double beam spectrophotometer (q.v.).

Double beam spectrophotometer: instrument where the light beam (of a single wavelength) incident on the sample is split, with half the light passing through a sample and the other half passing through a reference cell. The instrument is usually set up to record the difference in absorbance between the sample and reference beams. Sometimes known as a split beam spectrophotometer.

Dual wavelength spectrophotometer: instrument where a single sample is illuminated alternately (with a rapid alternation) by light of two different wavelengths, and both absorbances recorded (commonly as the difference between them). A dual wavelength spectrophotometer requires two separate monochromators to determine the two illumination wavelengths.

Extinction coefficient: molar absorptivity.

Fluorescence quencher: compound causing a decrease in emitted light from an excited fluorophore by providing an alternative pathway for loss of energy from the excited state. Fluorescence quenchers decrease the quantum yield (q.v.) of a fluorophore. Water and O_2 are common quenchers in biological samples.

Fluorophore: that atomic grouping within a molecule or system responsible for fluorescence emission.

Isosbestic point: that point in a spectrum that remains unchanged during a chemical (or physical) transition. The point where a difference spectrum (q.v.) crosses the abscissa. There is always at least one isosbestic point in spectra taken before and after a transition between two states.

Molar absorptivity (ε_λ): absorbance at wavelength λ of a 1 M (1 mol l^{-1}) solution over a path length of 1 cm.

Natural bandwidth: width at half height (in nm) of the peak of sample absorption or fluorescence.

Nucleotide: molecule of the general structure [heterocyclic base]-[sugar]-[phosphate]$_n$, similar to the monomeric components of DNA. The term includes the components of DNA and RNA (AMP, dGTP etc.) but also includes flavin mononucleotide (FMN), nicotinamide adenine dinucleotide (NAD^+) and similar structures.

Optically clear: a solution showing no detectable turbidity (but possibly showing absorption of light).

Quantum yield: (in fluorescence measurements) total number of photons emitted/number of photons absorbed.

Ratio fluorometer: fluorometer where either the excitation or emitted light is recorded simultaneously at two different wavelengths, and the result expressed as a ratio of fluorescence at the two wavelengths. A ratio fluorometer requires three monochromators – two to provide the two emission (or the two excitation) wavelengths and a third to select the single wavelength for excitation (or emission).

Ratio spectrophotometer: instrument where a small fraction of light is diverted from the incident beam to fall on a reference photodetector. This allows correction for fluctuations in lamp output to be made to the recorded signal.

Resonance energy transfer: energy absorbed as a photon by one chromophore transferred to a second chromophore by direct excited state dipole: ground state dipole interaction (i.e. without re-emission of the energy as a photon). It requires a suitable spacing and alignment of the two chromophores.

Reverse optics: spectrophotometer design where white light is passed through the sample, and only subsequently dispersed (by a monochromator) for selection and measurement. In most spectrophotometers, the sample is illuminated by light of only a narrow band of wavelengths which has been selected by a monochromator between the lamp and the sample ('conventional optics').

Specific absorptivity ($\varepsilon^{1\%}$): absorbance of a 1% (10 g l^{-1}) solution over a path length of 1 cm.

Spectral bandwidth: central half of the band of wavelengths passed by the exit slit of the monochromator. It is determined by the dispersion of the monochromator and its exit slit width. The spectral bandwidth limits the resolution of a spectrophotometer; if the spectral band width is of the same order as the natural bandwidth (q.v.) of an absorption or fluorescence peak, the peak will appear distorted.

Stark effect: (red) shift of absorption peak when the chromophore is present in a strong electric field. Observed as the carotenoid shift (q.v) in photosynthetic systems.

Stray light: light reaching the detector, (a) other than that passing through the sample, or (b) at wavelengths other than those selected by the operator. Stray light is the major instrument-related source of error in spectrophotometric measurements, and becomes more significant as absorbance increases.

Transmittance: the fraction of incident light passing through a sample (I / I_0). Absorbance = – log transmittance.

Turbidity: a decrease in light transmitted through a solution due to removal of energy by scattering by particulate material in the light path. The size of particles giving rise to turbidity must be of the order of magnitude of the wavelength of illumination.

Appendix B

Suppliers

Applied Photophysics Ltd, 203–205 Kingston Road, Leatherhead, Surrey KT22 7PB, UK. Tel. (0) 1372 386537; Fax (0) 1372 386471.

ATI Unicam Ltd, York Street, Cambridge, Cambs CB1 2PX, UK. Tel. (0) 1223 358866; Fax (0) 1223 312764.

Beckman Instruments (UK) Ltd, Oakley Court, Kingsmead Business Park, London Road, High Wycombe, Bucks HP11 1JU, UK. Tel. (0) 1494 442233; Fax (0) 1494 463836.

Boehringer Mannheim (UK) Diagnostics and Biochemicals Ltd, Bell Lane, Lewes, East Sussex BN7 1LG, UK. Tel. (0) 1273 480444; (0) 1273 480226; Fax (0) 1273 480266.

Calbiochem-Novabiochem (UK) Ltd, Boulevard Industrial Park, Padge Road, Beeston, Nottingham NG9 2JR, UK. Tel. (0) 115 943084; Fax (0) 115 9430951.

Cecil Instruments Ltd, Milton Technical Centre, Cambridge Road, Milton, Cambs CB4 6AZ, UK. Tel. (0) 1223 420821; Fax (0) 1223 420475.

Centronics Ltd, Centronic House, King Henry's Drive, New Addington, Croydon, Surrey CR9 0BG, UK.

Fisher Scientific Equipment, Bishop Meadow Road, Loughborough, Leics LE11 0RG, UK. Tel. (0) 1509 231166; Fax (0) 1509 231893.

Hellma (England) Ltd, 23 Station Road, Westcliff-on-Sea, Essex SS0 7RA, UK. Tel. (0) 1702 335266; Fax (0) 1702 430652.

Hereus Quartzschmelze GmbH, D-6450, Hanau, Germany.

Hi-Tech Scientific, Brunel Road, Salisbury SP2 7PU, UK. Tel. (0) 1722 323643; Fax (0) 1722 412153.

Hitachi Scientific Instruments, Hogwood Industrial Estate, Finchampstead, Wokingham, Berks RG11 4QQ, UK. Tel. (0) 1734 328682; Fax (0) 1734 328622.

Kontron Instruments Ltd, Blackmoor Lane, Croxley Business Park, Watford, Herts WD1 8XQ, UK. Tel. (0) 1923 245991; Fax (0) 1923 412211.

Light Path Optical Co. Ltd, Unit 3, Elms Industrial Estate, Church Road, Harold Wood, Essex RM3 0JU, UK. Tel. (0) 1708 349136; Fax (0) 1708 381638.

Molecular Devices, 1311 Orleans Drive, Sunnyvale, CA 94089, USA. Tel. 408 7471700; Fax 415 3222069.

Molecular Probes, Inc., P.O. Box 22010, Eugene, OR 97402-0414, USA. Tel. 541 4658338; Fax 514 3446504.

Oriel Instruments, 250 Long Beach Boulevard, P.O. Box. 872, Stratford, CT 06497, USA. Tel. 203 3778282.

Perkin-Elmer Ltd, Post Office Lane, Beaconsfield, Bucks HP9 1QA, UK. Tel. (0) 1494 676161; Fax (0) 1494 679331.

Pharmacia LKB Biochrom Ltd, Science Park, Milton Road, Cambridge CB4 4FJ, UK. Tel. (0) 1223 423723; Fax (0) 1223 420164.

Photon Technology International (PTI), Suite 3, The Sanctuary, Oakhill Grove, Surbiton KT6 6DU, UK. Tel. (0) 181 3906676; Fax (0) 181 3906610.

Shimadzu Europa GmbH, Albert-Hahn Strasse 6-10, D-47269 Duisburg 29, Germany. Tel. 49 203 76870; Fax 49 203 7666.

SLM Instruments Inc., 810 W. Anthony Drive, Urbana, IL 61801, USA. Tel. 217 384 7730; Fax 217 384 7744.

spex Industries, 3880 Park Avenue, Edison, NJ 08820, USA. Tel. 908 549 2380.

Thorn EMI Electron Tubes Ltd, Bury Street, Ruislip, Middlesex HA4 7TA, UK. Tel. (0) 181 6062500; Fax (0) 181 6062517.

Varian Ltd, 28 Manor Rd, Walton-on-Thames, Surrey KT12 2QF, UK. Tel. (0) 1932 898000; Fax (0) 1932 228769.

Appendix C

Further reading

Atkins, P.W. (1982) *Physical Chemistry*, 2nd edn. Oxford University Press, Oxford.

Ball, J.E. (ed.) (1981) *Spectroscopy in Biochemistry*. CRC Press, Boca Raton, FL.

Bergmeyer, H.U. (ed.) (1974) *Methods of Enzymatic Analysis*. Verlag Chemie, Germany.

Burgess, C. and Knowles, A. (1981) *Standards in Absorption Spectrometry*. Chapman & Hall, London.

Campbell, I.D. and Dwek, R.A. (1984) *Biological Spectroscopy*. Benjamin/Cummings, Menlo Park, CA.

Freifelder, D. (1982) *Physical Biochemistry,* 2nd edn. Freeman, New York.

Hadfield, A. and Hajdu, J. (1993) *J. Appl. Crystallog.*, **26,** 839–842.

Harris, D.A. and Bashford, C.L. (eds) (1987) *Spectrophotometry and Spectrofluorimetry: a Practical Approach*. IRL Press, Oxford.

Haugland, R.P. (1992) *Handbook of Fluorescent Probes and Research Chemicals*, 5th edn. Molecular Probes, Inc., Eugene, FL.

Hipkins, M.F. and Baker, N.R. (eds) (1986) *Photosynthesis Energy Transduction: a Practical Approach*. IRL Press, Oxford.

Hirs, C.H.W. (ed.) (1967) *Methods in Enzymology,* Vol. 11. Academic Press, New York.

McCormack, J.G. and Cobbold, P.H. (eds) (1991) *Cellular Calcium: a Practical Approach*. IRL Press, Oxford.

Miller, J.N. (1981) *Standards in Fluorescence Spectrometry*. Chapman & Hall, London.

Tsein, R.Y. (1989) *Ann. Rev. Neurosci.*, **12,** 227–253.

INDEX

Milton Keynes UK
Ingram Content Group UK Ltd.
UKHW031150141024
449569UK00024B/920